The
CHIEF
of
STAFF

An Insider's Perspective
to Becoming a Strategic Partner
in the Executive Suite

ANNE MARIE OTANEZ

Special Dedication:

Je t'aime mama et papi. Pour toujours.

-Anne Marie

Acknowledgments:

Chris: When I told you I was on this journey, you simply supported me. I love you!

Serge: If everyone had a brother like you, they would only know joy, happiness, and success.

Dr Avis: Without you, none of this happens! And there is still more. Everything is possible in this reality.

Michael Parker: A chance encounter during a meeting has made all the difference in the world for me.

Elizabeth: We are doing big things. Thank you for inspiring me.

Kim: What can I say? A fierce warrior!

Azza: ma belle, MA belle!

My board of directors: Those who have offered me mentorship and sponsorship — you listened to me and provided me with critical feedback. Those opinions helped shape my journey. Those names would fill a book on their own...

Desireé: As my editor, you made this process so easy. I never thought I would pen a book and here I am. Your feedback was invaluable, and I appreciate the time and effort you put into improving my piece. Wow... author! Thank you again, Desireé.

Tom, Victoria, Jay, Rebekah, Will, Andrea, Ryan, Don, Joe, Tana, Tammi, and too many more to name.

MESSAGE FROM THE AUTHOR

As a chief of staff, I have encountered numerous challenges and experiences that have shaped my professional and personal growth. Your support means the world to me, extending beyond a mere transaction and representing a shared connection and mutual appreciation for the complexities and nuances of leadership.

I am truly humbled by your decision to embark on this literary journey with me, and I am honored to have you as a reader. Readers like you make the effort, time, and passion invested in writing this book worthwhile. I am grateful for the opportunity to connect with individuals who share a common interest in leadership, strategy, and the intricacies of the chief of staff role.

Creating this book has been a labor of love; knowing it has found its way into your hands brings me immense joy. Your decision to purchase this book validates my efforts and inspires me to continue sharing my knowledge and experiences with the world. Therefore, I want to express my sincere appreciation for your support in spreading the word about this book. While your recommendations, social media shares, and word-of-mouth endorsements have been instrumental in reaching a wider audience, your advocacy for this work helps me as an author—assisting others who can benefit from the knowledge and insights it contains.

I sincerely hope that the insights and lessons shared within the pages of this book resonate with you, offering inspiration and guidance in your endeavors. I wish you much inspiration and success on your leadership journey.

Thank you again!

Anne Marie
Chief of Staff
Staff and Technology-Operations Leadership

CONTENTS

FOREWORD

Ever wrestle with balancing product deadlines, sudden shifts in market demand, and keeping your team motivated? As a former military pilot, I know that feeling of having to be both a strategist and a firefighter. But in the tech sector, the pace is fast, and failure to adapt can cost you and your team. After a lifetime of military service, from acts of bravery to heartbreaking loss, I thought I'd seen it all—whether above the skies of Iraq or in the Neonatal Intensive Care Unit. Then I met Anne Marie.

When I took on my first tech CoS role, there was no body of knowledge. There were no roadmaps. There was no doctrine. Not another textbook but a seasoned mentor, showing me how to handle funding roadblocks, forge partnerships with skeptical engineering teams, and anticipate the next disruptive tech—all without sacrificing team morale. That's the kind of guide I found in Anne Marie. In the relentless battleground of the tech sector, I found a civilian "battle buddy" whose leadership and compassion rivaled any commander I'd served under.

This book embodies her unique brilliance. It became my roadmap to transition from a seasoned veteran to a respected innovator in the tech world. Anne Marie showed me new depths of the power of collaboration, the importance of staying ahead of the curve, and a new way to inspire those around me. Our partnership, where my military precision met her tech-savvy leadership, shattered expectations. Her guidance became my North Star, ensuring we consistently exceeded the goals we set.

If you're a chief of staff ready to level up or a tech leader looking to empower your right hand, this book is your field manual. *The Chief of Staff: An Insider's Perspective to Becoming a Strategic Partner in the Executive Suite,* shaped by Anne Marie's hands-on experience, is that guide for YOU. Within the pages that follow, you'll gain the strategies that propelled me from tactical firefighting to proactive innovation, and you'll discover transformative insights that are not only theoretical but also practical and applicable to your role as a tech leader or chief. It will equip you with the foresight needed to navigate this fast-paced environment, promising personal growth and professional success.

Prepare to dive in deep. Engage with the material, and you'll emerge a tactical innovator ready to tackle the biggest tech challenges with confidence and impact.

Elizabeth Stephens
CEO | DBS Cyber
United States Marine Corps Major, *Ret.*

INTRODUCTION

I'm the chief of staff. Sounds cool, right? Some folks will look at you with a bit of awe because it's a C-title. Others are baffled at the term. But, what exactly does the chief of staff do, and why would you want to be one?

The questions above are why I wanted to write this book. I want to capture and share my unique perspective, ideas, and experiences with the world about being a chief of staff (CoS). My goal is to educate and contribute to the ongoing conversation of what a CoS does and how you can be impactful in this role. During my tenure, I have spent time educating people I work with, mentors, and sponsors to have a well-rounded view of what the job entails and how to maximize success while enjoying job satisfaction during the broccoli moments. What's a broccoli moment, you might ask? This is a fun analogy that will help you through this reference. I have not always been a fan of broccoli, but I've come to like it. Some aspects of your job you will love, and some aspects may not be your favorite, but your broccoli moment is your ability to learn something new or different that you may not necessarily have thought to lean into.

"Chiefing" is akin to a person at a buffet: Like many buffets, the food seems endless. Similar to a chief, there are multiple ways to perform the job function, and sometimes, two chiefs at the same company and on the same team have very different success measures. At a buffet, you find some staple items everywhere, i.e., your salad section will always have lettuce and greens and a mixture of a dozen or more items on the menu. Those staples will be some of the core skills every chief of staff needs regardless of their role, for example, a measure of ghostwriting for your executive in their voice.

A CoS is a force multiplier who navigates and deals with challenges during their tenure. In the tech world, ambiguity is often the only constant, as many will come to understand. They must be comfortable operating in an atmosphere where conditions can change rapidly and little direction may be provided. Adapting to unpredicted challenges and making informed decisions in the face of ambiguity is a skill that defines success in this role.

Another challenge the CoS faces is maneuvering the world of interpersonal subtitles and office politics, so building solid relationships with both technical and non-technical stakeholders is crucial for success. Being a trusted advisor requires strategic dexterity and the ability to navigate the many cultures and communities within the organization, recognizing the value each brings to elevating the diverse side of the organization.

A staff-related issue was one of the first things I helped shape when I began as chief of staff. My executive and I had worked previously and established a working relationship. Within my first month, there was a restructuring, and one particular person, we will call Julie, was potentially in a position to move into a leadership role. I had only been on the team for a short while, and even in my newness, several employees asked for some time to have a confidential conversation about her ability to work across teams. After reaching out to several of my mentors for help and perspectives, I was able to have a challenging conversation with my leader about Julie, a person I had not previously interacted with.

In the end, I was not making a decision about Julie and her ability to move into a leadership role. The discussion with my leader was about ensuring she had enough data points regarding Julie's management style, her relationship with her peers, and the feedback received. They then had a more extended conversation about many other topics, which helped Julie to land in a good spot.

Now, you may think this example was about Julie. It was not. It was about the ability of a chief of staff to have an immediate impact and influence. I needed the confidence to have a critical conversation with my leader while providing balanced information. The discussions about Julie and her growth had already been happening, and any questions about her ability to manage had previously been discussed. However, this was an opportunity to think through my narrative, gain credibility with the team, and be a partner to my leader. The net is that you will have hard conversations as you delve into this world and opportunities to see what skills necessitate success.

There are many ways to become a chief of staff. The traditional method is applying for a relevant position, interviewing, and being selected as the best candidate for various reasons. Another route is the familiar; knowing the hiring manager will help you land the role. I am a career program/project manager who transitioned into the CoS role. I was responsible for planning, executing, and closing organizational projects—a central point of contact, coordinating efforts among team members, stakeholders, and resources to ensure successful project delivery. Effective project managers possess strong leadership, communication, and organizational skills to navigate challenges and meet project goals. They manage teams, sit in with the executive to provide status and recommendations, and, if something breaks, figure out how to fix it. It can seem like a thankless job, but quite the contrary. I worked with a vast array of individuals, sharpened my strategic chops, expanded my knowledge about the area I was focused on, moved quickly when issues arose, and survived the frying pan heat when all went wrong. It was exciting, demanding, and refining.

Being a chief of staff leverages all of my previous experience. However, you don't have to have been a project manager to move into this role; various roles can be predecessors to a CoS role as they often require individuals with diverse skills and experiences. I have many peers with backgrounds in communications, sales, software engineering, and other fields. Typical roles that can serve as stepping stones to this position include Operations Manager, Project Manager, Program Manager, or Director of Strategy. These roles typically involve responsibilities related to executive support, strategic planning, and operational oversight, providing a foundation for the multifaceted responsibilities of a chief of staff. Many terms are associated with this title: liaison, trusted advisor, backup up, and proxy. And, as you review the following chapters, I will provide a behind-the-scenes perspective on what is essential to your function as a CoS—especially for those working in the technical industry, where you might see a variety of titles such as business manager, business program manager, director of operations, General Manager, or Chief Operations Officer (COO).

Understand that the CoS fills a role in the organization determined by the executive being supported. Your role will be vital to successfully maintaining your team and organization and creating a community and culture for the team. You will be in a position to help strategize the direction of the business while anticipating potential roadblocks and work with either your team if you have direct reports, the peers who sit with you on the leadership team, and, when possible, a matrixed team, to come up with solutions or the reason for why the outcome was different than predicted. Dealing with uncertainty and ambiguity will be just another day as you work through the day and keep things running smoothly. You will find that you are not asking what to do, but you have a list of things to do, and your leadership team will look to you for guidance, help, and where their focus should be.

Today, I am an executive technologist and business leader specializing in program and project management and the chief of staff development. I am a speaker, coach, mentor, and executive trainer. With over twenty years of experience working in the Hi-Tech, Healthcare, Automotive, and Entertainment industries, I bring a wealth of knowledge and expertise in staff and technology operations leadership. I not only worked for established companies but also worked the rigor of a few start-ups.

I have successfully driven key initiatives with a worldwide impact, including creating D&I Programs, Chief of Staff Academies, and leading key partnership deals. Being a CoS offers the gratification of orchestrating efficient operations within an organization, fostering collaboration among different departments, and playing a pivotal role in decision-making processes. The role brings the rewards of strategic influence, leadership development, and the satisfaction of contributing to the overall success and cohesion of the team or company.

One of the first leadership offsites I ran helped me uncover a skill and passion for leadership development. I had been in this organization for several months and had a good feel for some areas of discussion. I created a small volunteer team and outlined some topics to discuss with potential facilitators. I also investigated a variety of leadership relationship-building exercises. I proposed a goal for the offsite and connected the topics to the goal. With this being my first offsite, my stress level was high. My hope was for connection and intentional discussions about strategy. Once the offsite was done, I asked for feedback, which was overwhelmingly positive, with minor tweaks for future offsites. It felt good, and I was excited to implement the tweaks.

Nevertheless, not all things go well, and this role is fraught with challenges, as mentioned earlier in my example about Julie. Those critical conversations can be difficult and bring a different type of stress, and sometimes, there is not always a "happy ending." This reminds me of a Nelson Mandela quote: "I never lose; I either win or learn." There will be plenty of opportunities to discover and grow. You will win some, and you will learn a lot. Thus, remain open so you can either win... or learn.

So, let's start this journey together: what a chief is, what a chief "ain't," and how do I do it and find success?

ANNE MARIE OTANEZ

1

WHAT IS A CHIEF EXACTLY?

The chief of staff is the person who makes sure that everything works smoothly. They are usually the first to show up theoretically and the last to leave. An excellent analogy for a CoS is a conductor to an orchestra; the writer of the musical score is your executive, and you ensure that each person/team is playing the right part to create harmonious pieces.

As with conducting, it will take a moment before each person learns and understands their part, but once that is achieved, the outcome is something everyone can be proud of. The fundamental point is that each part is essential to the result, from the smallest instrument to the loudest drum. There are many roles to take on. We will talk about a few of these, and as you continue to build skills, you will add more pieces to that puzzle.

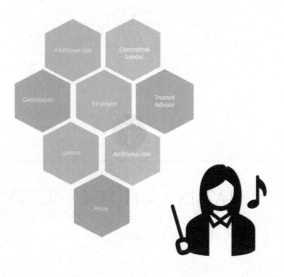

The chief of staff is a trusted advisor and thought partner to the executive they support. You will find that you are a listening ear or sounding board and should have a good view of the company to provide solid advice, comments, and recommendations for your executive. You have an integrated understanding of how the organization is connected from the leadership perspective to the many pieces deep within each team—knowing how the puzzle pieces fit.

A chief of staff is a trusted advisor, a gatekeeper, a strategist, an operational leader, and a liaison.

A Trusted Advisor is an individual who works in tandem with their executive, mainly as a strategic partner. Harnessing extensive knowledge and experience, trusted advisors provide advice and tailored guidance. As a trusted advisor, you'll be sitting in meetings that are confidential, finding that some of your partners with leadership may not be privy to some of these conversations. It is your job to comprehend all of the variables in order to make informed choices. That executive needs to know they can come to you with anything, and you will keep the information confidential while also working to be unbiased in how you proceed.

I have sat in meetings with my leader where a reorganization was needed. We partnered with the human resources manager and the legal teams. When a team or company reorgs, oftentimes, roles are eliminated, and job scopes change. These discussions occur months in advance. As a trusted advisor, you work with some individuals who may be impacted, and that information cannot be shared. I often need to maintain confidentiality to safeguard sensitive information, ensure the trust of stakeholders, protect the organization's competitive advantage, and mitigate potential risks associated with the unauthorized disclosure of critical information. You will have an opportunity to develop executive maturity and understand the depth of conversations needed at various levels within the organization. You are strategizing and executing the vision.

An article published by Harvard1[1] states that the key principles for building trust as a consultant or advisor include demonstrating low self-orientation, prioritizing client needs over personal interests, asking probing questions to understand the client's perspective, providing options in a non-confrontational manner, admitting when there is uncertainty, being generous with knowledge sharing, fostering close communication and relationships with clients, building reliability over time by consistently delivering on promises, and actively practicing the skill of active listening to discern both spoken and unspoken client concerns. These principles, collectively, contribute to establishing credibility, transparency, and a strong foundation for long-term client trust. These eight simple tips will help you build

---

[1] https://www.harvard.co.uk/eight-steps-becoming-trusted-advisor/

the skills to earn the trust and respect of your peers and establish your reputation as a problem solver and person with critical insights into the business and organization.

The CoS is also known as **a Gatekeeper**, which can be challenging because many people in your organization may want an audience with the leader(s) you support. Some individuals require time to provide an update, and others may be looking for assistance or help. There are even others looking for mentoring or development opportunities to be better connected with the business.

These are only a few reasons your executives and leadership bodies are connected to the business. This is why it is vital to develop your rhythms of connection, which identify the listening rhythms for people to interact with each other and the leadership team. As you gain more information as a trusted advisor, you will also expand your insight into the business. You will be able to help your executives triage and determine which conversations need to be elevated to them and which ones require a different audience.

In the beginning, this will take some time. Your executive will also find themselves in a position when there are stakeholders outside of those they need to connect with, and you want to ensure that you provide the time and the space for those conversations. But with regular communication and an understanding of the needs of the business, you'll find yourself with a heightened ability to know what needs to be prioritized while ensuring that all the proper conversations are being had. You will be in a position to not only say yes but also "NO" as you protect your executive's time.

An empowered chief of staff controls the flow of people and documents; therefore, a gatekeeper has a critical function in small or large businesses. They are the person who controls access, either to someone in authority or to information—whether a message or a visitor will reach the decision-maker or not. This is an essential function because you are making sure that executives get what they need to operate in a well-informed manner while mitigating interruptions from visitors or callers. As you grow your operational team, this function can be delegated to a direct report on your team. On the flip side, a weak or unsupported CoS cannot ensure equity of all departments are considered nor that the staff produces consistent external messaging.

As a **Strategist**, you partner with your executive and leadership body to help move the plan to fruition. The leadership body will create its team strategies and vision leading up to your executive's overarching vision for the organization. You are managing the intricacy of successfully running a business.

Sometimes, a marketing officer's and a strategist's roles may seem similar since they are both crucial to creating and developing strategies based on what is happening in the industry to secure a competitive advantage and generate demand. Where a marketing executive works to grow the business through branding, converting customers, and promoting brand awareness, the strategist focuses on organizational, revenue, and financial changes that could improve the company's profits and market share. You are setting goals, creating tasks and initiatives, ensuring a path to completion for the plan, and providing needed updates when the plan changes.

Some strategists will provide dashboards, scorecards, and a variety of metrics to analyze and determine the plan's health. While typical dashboards may provide green, yellow, and red indicators, others may provide trending information to gauge movement and direction. There are a variety of methods and tools that can be used. Some data may be housed in data warehouses, and automated reports are generated. These reports can be created or modified to meet your business needs. As you work with your executive, you will

66

A strategist analyses
complex environments
or problems and designs
practical pathways and
business solutions to
achieve organisational
objectives. A strategist
is someone who has the
ability to see beyond
the near term.

-KAYE GLAMUZINA,
HEAD OF STRATEGY

determine which metrics will inform your org on the health of the business and the frequency of those reports.

Reports and dashboards are essential for many reasons. They provide you with a method to share information in your company that is easy to take in, offering insights that can be acted upon, giving the right narrative to the correct stakeholder to drive business performance, and providing a solid view into performance across the various teams with visuals and color coding for ease of reading and understand the message that is being communicated. Effective reports provide source and data information for those users who want to delve further into the details. The best thing about the reports from the CoS office is that you are the source of record; therefore, your data and information must be well vetted.

As an **Operational Leader**, you will create processes, procedures, and guidelines. This is highlighted later in this book under core skills. Whether a new or existing business or a part of a larger organization that requires integration with another team or an independent organization, you will need processes to ensure this is done with some level of predictability, consistency, and uniformity. A chief of staff supports their executive leader in many capacities. They typically supervise and communicate with lower-level staff members, provide project management, and implement strategic planning processes.

As a chief of operations, you are responsible for monitoring the company's overall operational process, maintaining efficient project management, and ensuring accurate project deliverables. You are responsible for ensuring that your executive is prepared for meetings and managing their communications. As a leadership team member, you will partner and lead initiatives with your peers and senior levels of the organization, working across departments to align teams, set goals, execute initiatives, and improve processes. This aspect of the job is operational, focusing on the company's day-to-day management and overseeing the execution of business strategies. You will help teams collaborate toward a singular vision—taking in different perspectives and finding a mutually shared purpose to establish a well-thought-out and inclusive foundation.

One of the leaders I partnered with found herself in many meetings, and as I worked with her administrator, we reviewed her schedule on a regular basis. On several occasions, she would reach out to me and ask me for details for an upcoming meeting. Because I had been regularly reviewing her calendar, I would have the details on hand, and it would be easier to prepare her for the meeting.

The Liaison is the executive's representative in many forums. I have sat in executive meetings for my leaders as their representatives when they cannot attend. You are able to speak for and characterize their perspectives; you are the champion for your executive. As a liaison, you are also the person your executive's key stakeholders turn to for assistance in the absence of your executive. The chief is a partner with the executive and a partner in identifying and executing the priorities and strategies. There are many critical key relationships a CoS will maintain, and some of these can be direct reporting relationships as part of the office.

There is an ongoing discussion about the role of an executive assistant and a chief of staff. While both roles enable leaders to do their best work, the CoS role differs from that of an EA. Unlike an executive assistant, a chief of staff works autonomously and does not handle routine correspondence or manage the leader's day-to-day schedule. However, the chief and assistant work in tandem to handle maintaining, running, and executing against the organization's visions in a strategic and transactional mode.

Still, the role of a chief is dynamic, requiring a versatile set of personality traits to effectively navigate and support the organization's leadership. While individual characteristics may vary, some common personality traits include the following:

Flexibility: Chiefs of Staff often work in fast-paced and ever-changing environments. An ability to adapt quickly to new situations, challenges, and priorities is crucial.

Communication Skills: Strong communication skills, both written and verbal, are vital. Chiefs of Staff often act as intermediaries, conveying information between departments and the leadership team.

Credibility: As the right-hand person to top executives, Chiefs of Staff must be trustworthy and discreet. They often handle sensitive information and need to maintain confidentiality.

Problem-Solver: The ability to identify issues, analyze problems, and propose effective solutions is valuable. Chiefs of Staff play a role in addressing challenges and streamlining operations.

Detail-Oriented: Chiefs of Staff often deal with intricate tasks and must ensure nothing falls through the cracks. Therefore, being meticulous and detail-oriented is crucial for overseeing and managing complex projects.

Time Management: Chiefs of Staff are typically involved in various tasks and responsibilities. Effective time management skills help them prioritize and accomplish multiple objectives efficiently.

Creativity: Chiefs of Staff often work with limited resources and need to find creative solutions. Resourcefulness allows them to make the most of available tools and opportunities.

Collaborative Mindset: Collaboration is critical to the role of a Chief of Staff. They work with diverse teams and departments, requiring a collaborative mindset to foster cooperation and unity.

Emotional Intelligence: Understanding and managing emotions, both their own and those of others, is crucial. Emotional intelligence helps Chiefs of Staff navigate interpersonal relationships and handle delicate situations effectively.

It's important to note that these traits may vary based on the specific needs and culture of the organization. Chiefs of Staff often tailor their approach to match the leadership style and requirements of the executives they support.

While a chief may focus on a specific area, they are not limited to that domain; some will be heavily involved in the communications aspects of their executives—deeply involved in their social and media presence. They may write or have individuals or vendors on their team to craft, schedule, and maintain that presence, engaging with target audiences. This will help establish and build credibility in the industry. This also will help establish your executive as a voice and leader in the industry, eventually making them a subject matter expert (SME). Your executive will have opportunities to speak in various capacities.

This role drives and coordinates the communication activities of an organization. They are tasked with developing and implementing communication strategies aligning with the organization's goals and values. They curate and distribute all communication materials, such as press releases, newsletters, brochures, social media posts, website content, etc. Depending on the visibility of your executive and as they are recognized as a leader in the industry, they may have connections in the media, and those relations should be maintained. That includes representation from media outlets, journalists, influencers, and other stakeholders. Using data and feedback, they constantly scrutinize and evaluate the impact and effectiveness of communication campaigns and initiatives. In those cases where issues arise, they manage and resolve any communication issues or crises that may occur, such as negative publicity, customer complaints, or internal conflicts.

Let's consider a scenario where a chief with communications expertise is faced with a challenge related to a potential reputational crisis, and the company has encountered an unforeseen issue that could negatively impact its public image: The CoS manages the communication strategy to mitigate the fallout and maintain the company's reputation, immediately engaging with the communications team to assess the situation, understand the scope of the issue, and identify key stakeholders.

Crafting a clear and transparent message becomes crucial, balancing the need for openness with protecting sensitive information. The CoS collaborates with the CEO, the legal team, and other relevant departments to ensure a unified and consistent response. They may also coordinate with external communication consultants or PR experts to develop a strategic plan for addressing media inquiries and managing the narrative.

In this situation, the chief leverages their communications expertise not only to navigate the crisis effectively but also to proactively communicate the company's commitment to transparency and resolution. This helps safeguard the company's reputation and stakeholder trust while ensuring that all communication activities adhere to the organization's brand identity, style guidelines, and ethical standards.

Other chiefs will be deeply entrenched in the sales domain. Sales excellence is a term that refers to the ability of a sales organization to achieve and exceed its growth objectives consistently. It involves a combination of factors, such as skills, knowledge, tools, training, content, culture, and data, that enable sales reps to deliver value to every customer interaction and optimize their sales performance, productivity, and proficiency. Depending on their industry, goals, and strategies, different businesses may have other definitions and measurements of sales excellence.

To have an idea of the key metrics or key performance indicators (KPIs) that can be used to assess sales excellence are:

> **Total revenue:** The amount of money generated by sales activities during any period. Measurements can include daily changes, week-of-week (WOW), month-to-date (MTD), Quarter-by-Quarter, Year-to-Date (YTD), Year-over-Year (YoY) growth, and other increments to highlight the picture.

Percentage of sales reps attaining 100% quota: The proportion of sales reps who meet or surpass their assigned sales targets in a given period. Depending on the size of your team, this will help to identify people who meet or exceed the pipeline requirements and their ability to hit those targets. This is an important view because having aggressive goals requires the staff to attain them. While this seems obvious, some organizations do not always have a sales team to match their targets, exerting extreme pressure on a sales team. Ensure you understand what this balance needs to be to achieve your target.

I can recall when I was working with one of our sales managers. He wanted to ensure the sales reps were aligned in how they approached potential customers. As we discussed some options, we decided to embark on sales team training for his team. We had reviewed a few areas he wanted to target and identified the type of training needed. We secured a location and a date, and this manager designed the curriculum for his team.

During the two-day meeting, he reviewed the current state and the type of customer to target. They practiced scenarios with potential outcomes and reviewed current offers and how they may need to be adjusted based on the proposed outcomes. He was able to relay some relevant information and train his team. While I was not directly involved in the training, I partnered with them to facilitate this training of his team.

Percentage of revenue from existing customers: The income received from repeat or loyal customers during any given period. As you review this number, your teams will key on movement. Is this number stable with incremental growth, or does it spike up or down? New businesses will have a relatively small pool of existing customers that will help build your business's foundational base. As with an existing base, there is a focus on acquiring new customers. Your sales teams will also monitor the proportion of revenue that comes from acquiring new customers during a period of time. Also considered in this "new customer" are potential customers who have expressed interest in your product or service and are likely to convert into buyers.

Various activities provide context into how much time is spent on core sales tasks, such as prospecting, qualifying, presenting, negotiating, and closing deals. This includes the time sales team members spend searching for customers as well as the amount of time that sales reps spend learning and absorbing new information and skills from various sources, such as training courses, coaching sessions, webinars, podcasts, etc.

Time to first deal: The amount of time it takes a new sales rep to close their first deal. These metrics can help sales leaders and managers monitor and evaluate the effectiveness of their sales strategy, processes, and resources and identify improvement and best practices. Using data and insights to continuously redefine and enhance their sales excellence, sales organizations can gain a competitive edge and achieve sustainable growth. As discussed earlier, this data will typically be found in a dashboard and frequently a scorecard. The scorecards are maintained and reviewed continuously to gauge where movement is and trajectory.

The picture that this data will share is where there is steady growth, opportunities areas, and where the business is struggling and, depending on the data, the "why." There are a variety of tools that will connect and help track this information. Your organization may have proprietary tools or shelf-bought tools to keep all the information in one place and accessible.

Some teams need to focus heavily on an operations manager. This person oversees the running of the business, which involves providing data to the team. This can include a variety of reports for the leadership and the organization. They will also oversee staff and team concerns, elevating those items to management with actions for resolution.

I enjoy the operational aspect of this role. One of the resources developed was "NTK (Need to Know) Emails." And that is precisely what they were: information communicated in executive operations forums that must be shared with the various teams. They included a static header and date. I also listed the target reader. The content was usually a few paragraphs or less, and they highlighted the information I was sharing, whether any action was required, and who to contact for additional information.

One NTK email referenced the policy and update and covered an upcoming policy change. The effective date of the policy change was listed as the reason for the change. Those details were also listed on an internal website for future reference. It was a great tool for concisely delivering information to a large group.

Earlier, we referenced the many roles a Chief of Staff takes on. While some may narrow down on a specific role or duty, effective CoS offices contain some flavor of all of these.

QUICK TIPS

TIP #1: As an executive who wears many hats, take time to understand what areas need immediate attention and prioritize those with your executive with a target completion date.

TIP #2: Listen, Listen, Listen. You will spend time in a management role, so take a moment to listen more than talk.

TIP #3: Your development as a chief is important. As you settle into your role, be intentional about the skills you want to work with your leader.

2

CORE SKILLS

⌒ℯℓℓℯ⌒

I n 2020, I partnered with a fellow chief of staff and facilitated a training program. We worked in the same industry but were on different teams and had solid CoS experience. We had discussed that there was no general chief of staff training and that it was needed. My colleague and I had multiple people in our organization ask us questions about how to be a chief, what the job entails, and other details about the role they were inquiring about.

We decided to facilitate a *Chief of Staff Academy,* partnering on an initiative and soliciting assistance from subject matter experts (SMEs) in specific areas. For example, since the CoS needs strong communication skills, we worked with a communications manager to provide a perspective, and we overlaid the chief of staff touch. Then, after reviewing our combined experiences, we identified these core skills that every chief should have.

This list is not comprehensive; however, it can serve as an excellent starting point to build your foundation. You will have experienced these skills in some manner, but as a chief, you will delve deeper into these areas.

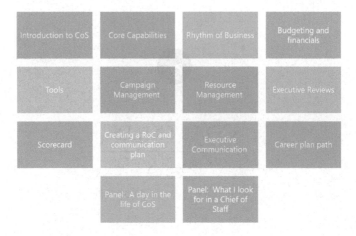

Introduction to CoS: This is really a summary to the role. These are the details you will find in this book as you either contemplate a role as a CoS or further strengthen your skills.

Core Capabilities: A CoS will have a mixture of leadership experience and acumen. You will have to deal with senior and executive teams, and understanding how to navigate the narratives needed at this level is vital to your success. You will handle escalations and issues that require immediate remediation with a plan.

> **Critical thinking**: Critical thinking isn't about being constantly negative or critical of everything. It's about objectivity and having an open, inquisitive mind. When you think critically, you use the data or information (not personal opinions, biases, etc.) to craft a solid sense of the state of affairs. Once you are in a place where you have a strong understanding,

you can make better decisions and solve problems more effectively.

Critical thinking is analyzing information and effectively forming an unbiased conclusion. To think critically, you must be aware of your biases and assumptions and apply consistent standards when reviewing sources. Critical thinking skills help you to:

- Identify reliable sources.
- Consider and provide a response to arguments.
- Review all potential possibilities.
- Test hypotheses against criteria.

Critical thinking has several benefits.[2] You will be empowered to make better decisions and solve problems better. If your goal is to become a persuasive communicator, this will put you on that track. One of the most significant benefits is your ability to cope with everyday problems as they arise while promoting independent thinking.

Project management is the process of leading the work of a team to achieve specific project objectives within agreed parameters. Project management involves planning and organizing a company's resources, such as personnel, finances, technology, and intellectual property, to deliver value to people. Project management can apply to a one-time or ongoing activity with final deliverables limited

---

[2] https://criticalthinkingacademy.net/index.php/ct/benefits-of-critical-thinking

to a specific timescale and budget. These skills will enable you to do this job well.

Successful companies recognize standards as business tools that should be managed alongside quality, safety, intellectual property, and environmental policies. Standardization leads to lower costs by reducing redundancy, minimizing errors or recalls, and reducing time to market. This *knowledge of business standards* for your industry is the input you will need as you lead and make decisions to run the operational aspect of your business and team.

Policy development is the process of creating, drafting, and enacting policies that address specific issues or problems. Policies are plans, positions, and guidelines that influence decisions by teams, organizations, or other bodies. Policy creation involves multiple stages, including identifying the issue(s), creating the actual policy, and implementing and evaluating the policy. Policies must be developed and guidelines established in an ever-changing working landscape. This has become more important as each industry is working to address a hybrid, remote, in-office environment while attracting and retaining employees from a variety of backgrounds.

Adaptability means to be flexible and accept change easily. Adaptable people change their behavior by evaluating the situation and adjusting to accommodate different conditions. Adaptable people thrive in challenging situations and see opportunities where others see failure. Being adaptable requires being innovative, open, and resilient — all of which will be beneficial in an ambiguous environment.

As a CoS, you will find vagueness in many aspects of the roles, and it's up to you to help clarify and minimize confusion.

One particularly interesting situation I experienced was when I was partnering with two executives. Our organization had decided to change the reporting structure, and instead of having two leaders of the org, one leader reported to the other. As you might imagine, this was a challenging situation. With the principal executive, we had a conversation to decide the best way to ensure that the new reporting structure would land most equitably while maintaining the fact that there was a new reporting structure. At the end of the day, the organization chose to fold one team into another.

Since it can be highly challenging, approaching a situation like this with respect and compassion can make the change a little less daunting. So, the three of us—the new primary leader, the secondary leader, and I—discussed the org as a whole, ways to effectively communicate the changes happening, and how to ensure that the momentum the team was experiencing would not be impacted.

Team management is the ability to lead a group of people to perform a task or achieve a common goal, which includes collaboration and communication while setting objectives. You will be involved in performance evaluations and other workforce management tasks as a manager, partnering with your team members as you problem-solve. Team management also requires supporting and motivating team members to work effectively and grow professionally. Finding opportunities for development and career growth will be an aspect of growing and scaling your team, organization, and business.

A **team management plan** may use strategies, methods, or software to help organize and manage the team.

Rhythm of Business (RoB): This method identifies and lists the key events, milestones, and activities scheduled across the business year, whether fiscal or calendar. It helps to manage teams better and achieve goals and objectives. Below is a sample of the flow for the rhythm of business.

Rhythm of Business

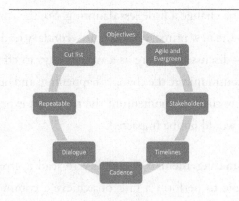

Your ROB will start with a purpose: What are you trying to accomplish? What are the goals and the needed outcomes? It needs to be agile and evergreen. Agile inspires constant change as your business adapts and scales to internal and external changes. Evergreen is in the name. It is the continuous growth through challenges.

Timelines will offer the when, how often, and how much time is needed, as well as the communication method — whether meetings, written communication, etc.—that occurs on some cadence. Details about timelines and communication methods are further elaborated in the communications section.

Now, let's talk about a cut list. What is it and why is it important? This is used to determine the prioritization of work to be completed, identifying the initiatives representing how a business will achieve profit. Organizations subtract above-the-line costs from their revenue to determine gross profit. Depending on the type of business and products or services they provide, the process for determining above-the-line costs may vary. Below-the-line initiatives may not immediately impact a company's profit-and-loss account. They are non-funded initiatives, so resources or time needed cannot be allocated to the current strategy.

A RoB is a governance model or the business cycle for operations, ensuring that employees never feel lost in the business cycle. All stakeholders provide a voice, encompassing the individuals and teams that need to be informed or responsible for the RoB.

Your RoB includes what is needed to facilitate complying with all the required HR processes and activities, providing visibility for finance processes and budget/activities such as reconciliation and trending. As you build strategic plans, various inputs, outputs, and artifacts will be created. These valuable pieces of information can all be found in your RoB.

If your team is part of a larger organization, it identifies milestones and events, helping you have a well-rounded RoB that will help your team understand what is expected and provide transparency and consistency.

Budgeting and Financials: Many types of budgets and financial information will require reporting. Tracking and reconciling the budget happen on a regular basis, as determined by your executive body. There are costs associated with special projects, customer engagement, traveling, product costs, cost of goods (COGS), training, hardware, team activities, and company swag, as well as HR-related costs associated with workforce planning (hiring, layoffs, job elimination, and retention). If you have revenue, you will track those costs.

Financial models and pricing calculations are also costs that will impact your reporting. Budgeting is the process of defining the direction your leadership team wants to move in. At the same time, financial reporting illustrates whether the company is reaching its budget goals and where it is heading in the future. Budget reports highlight different projected budgets throughout specific periods, including sales or operational budgets.

Financial reports show the results of your team's economic performance. Accuracy in this area can make or break your business. Quite simply, money in and money out should be reviewed often and adjustments made to hit revenue targets and KPIs. As trends reflect positive or negative directions, you can react quickly and appropriately.

Tools: The tools needed include any software or hardware requirements for your role. These tools may consist of access to specific programs. As a CoS, you may also need similar access that your executive has to run specific org reports. Understanding the required details or executive reports is the best way to determine what programs or databases you need to access. A team outside of your organization sometimes facilitates this, and there may be compliance or security requirements that may require attestation before getting access.

Campaign Management: Campaign management involves planning, executing, tracking, and analyzing a marketing campaign from start to finish. You will manage a campaign or a series of marketing activities to achieve a specific goal. This helps companies segment, target, and organize their multichannel campaigns. Successful campaign management starts with a successful marketing campaign, so here are the essential elements of a winning campaign:

- **Target audience:** This is the defined audience a company wants to reach with its marketing campaign. It should be a well-thought-out customer identity that will benefit from the offer.

- **Contact list:** A list of contacts that fit the target audience.

- **Value proposition:** Every successful campaign needs a good value proposition to drive the messaging and benefit. Your key stakeholders who are driving the campaigns will put themselves in the place of their customers to understand exactly how the product or service can solve their problems and concisely communicate this throughout the campaign.

- **Call-to-Action:** A CTA should provide potential customers with a reason to ask for more information while keeping them. It's commonly used in marketing and advertising to nudge potential customers towards a desired action, such as making a purchase, signing up for a service, or visiting a website. They should have a clear reason to purchase and use your product or engage with the services offered.

- **Delivery:** Campaigns need an action plan to deliver the promised message. This can be through emails, digital advertisements, or social media.

- **Follow-up:** Once the campaign has been launched, it is vital that you contact your customer again to answer any questions, respond to requests for help, and understand the next steps to secure that customer.

Resource Management: This can take on many different flavors. For a CoS, this includes working with the manager to hire and fire resources and ensuring that HR and recruiting play their roles in this part of the business. It also includes retention planning from a strategic perspective, where the teams may potentially grow. If you work for an international company, you may need to evaluate what aspects of the business must be developed in a remote location. This will also include planning for leadership growth and presence.

As you consider your resource management, you should ask what priorities you must deliver on to meet customer commitments, respond to emerging customer needs, or address competitive pressures. Be sure to consider immediate and longer-term needs and the overall impact on your resource allocations.

Executive Reviews: You will be involved in creating this content for your leadership team. These reviews are weekly, monthly, quarterly, and annual reviews. The audience can be cross-functional teams or executive leadership. Creating these reviews will require knowledge of the business goals and strategy and the method by which your organization will define success. The business reviews are comprehensive reports highlighting trends and showing how you line up against your competition. There are many types of reviews:

- Financial reporting
- OKR report out, KPIs updates, and status
- Status on various initiatives and goals
- If your business includes marketing and advertising campaigns, see how these campaigns are progressing and ask for help as needed.
- Customer feedback highlights company perception and how a product or program is perceived. It can also include customer satisfaction scores resulting from a survey of the customer.

Scorecarding: : A scorecard or dashboard is a business performance management tool. It is an organized report used to track how well or poorly the teams are performing against a specified set of activities.

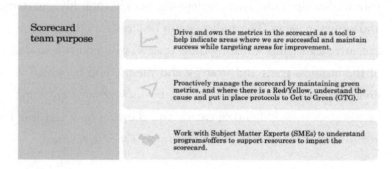

Scorecard team purpose

Drive and own the metrics in the scorecard as a tool to help indicate areas where we are successful and maintain success while targeting areas for improvement.

Proactively manage the scorecard by maintaining green metrics, and where there is a Red/Yellow, understand the cause and put in place protocols to Get to Green (GTG).

Work with Subject Matter Experts (SMEs) to understand programs/offers to support resources to impact the scorecard.

Executives use scorecards to gauge how "on track" things are and are usually focused on managing the application of your organization's strategy and associated operational activities. "In a 2020 survey, 88% of respondents reported using the balanced scorecard for strategy implementation management, and 63% for operational management. Although less common, the balanced scorecard is also used by individuals to track personal performance; only 17% of respondents in the survey reported using balanced scorecards in this way. However, it is clear from the same survey that a larger proportion (about 30%) use corporate balanced scorecard elements to inform personal goal setting and incentive calculations."[3]

Scorecards monitor and measure the progress of your team's metrics. Your scorecard will look at the past, present, and future with an eye for a few highlights, some areas for improvement, and an ask for help from management. It is another glance into how the business is doing and what areas require focus.

---

[3] https:2gc.eu/resources/survey-reports/2020-survey

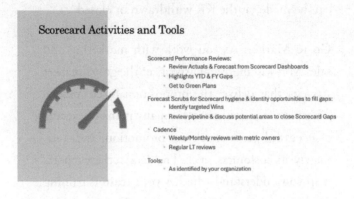

Scorecard Activities and Tools

Scorecard Performance Reviews:
- Review Actuals & Forecast from Scorecard Dashboards
- Highlights YTD & FY Gaps
- Get to Green Plans

Forecast Scrubs for Scorecard hygiene & identity opportunities to fill gaps:
- Identify targeted Wins
- Review pipeline & discuss potential areas to close Scorecard Gaps

- Cadence
  - Weekly/Monthly reviews with metric owners
  - Regular LT reviews

Tools:
- As identified by your organization

Here are a few examples of the types of reports created specifically for your team:

Recruiting: You will be tracking hiring activities with your recruiting staff. Items of interest will include open positions, candidates in the pipeline, time to hire from application and interview, and the number of candidates turned away with reasons listed in categories. In some cases, confidential information will include demographics, gender, race, veteran status, and disability status.

OKRs: You will track key objectives and their associated key results (KR) with status and blockers. In some cases, you will have KRs trending in a positive direction, and you can leverage and learn from the work being done and how the positive progress can be replicated. When the KR is not trending well, the reports are early indicators of a problem, and the root cause needs to be identified quickly so progress can be made, or the KR withdrawn or closed.

Go to Market: As you work with marketing and sales, you will highlight aspects of the go-to-market strategy that deliver the proposition to customers. This plan features how a company plans to execute a successful product release, promotion, and, ultimately, its customers' sales. The data in these reports help you understand whether your team is gaining or losing the competitive advantage being sought.

Creating an RoC and communication plan: A RoC is a Rhythm of Connection, which we introduced in the discussion about the communication plan. These rhythms identify each RoB (Rhythm of Business) and their place in the business.

But what is the difference between a RoC and a RoB? The RoC is the overarching plan that encapsulates all the of the RoB. You have one RoC with multiple RoBs, and the communication plan will provide additional levels of detail. This rhythm of connection model is sometimes called a governance or business cadence model. It doesn't need to be complex. In fact, the simpler it is, the better because you and everyone on your team will use it.

Having a joint RoB will allow the team to execute predictably and efficiently, reducing meeting overload and creating a shared sense of team. It will also minimize randomization, inefficiency, and siloed decision-making of separate RoB and forums. Ultimately, it ensures that our work addresses the established customer and business targets. Rhythm implies we have a sense of when something is supposed to happen and how often.

RoBs should have a defined update cycle and a rough calendar of when specific events are to occur in each cycle.

We all sigh when we look at our calendars, and someone cancels a meeting (happier when that meeting is not rescheduled). Proactively look at our meetings and recognize that this is not meeting our needs. Why do we have some meetings—can this be fulfilled by starting a dialogue in a chat (maturity level of using chat vs emails)? Too many meetings will lead to meeting fatigue. In the article published by HuffPost, stress levels spike when people are in too many back-to-back meetings. One recommended way to combat this is to walk away from your computer or device and take a walk, meditate, or spend a few minutes doing an activity that doesn't require using a device. Even checking email does not allow your mind time to reset and reengage effectively. Be deliberate about your breaks and use this time to take a breather.

The final step to your RoB is the tactical aspect. When should the meetings occur? Some teams opt to hold large, organization-wide meetings earlier in the week regularly so there is predictability. There will inevitably be conflicts. This will be a bit of a juggling act, but you can identify a few meetings that require a large group and keep those specific meetings on the same day. For example, Mondays can be for individual organizational meetings and Tuesdays for leadership and revenue meetings.

Check with the org leaders for days for cross-collaboration meetings and ensure that if you have teams that are not at the same time, the type of meeting is accounted for with times.

Executive Communication: This can involve several different facets. Your executive will need to communicate consistently with the team. The chief of staff may ghostwrite some of the communications. Just like you are prepping materials for your executive and gathering details to create review slides, you must also ensure regular messaging from the leadership body.

Written communications can include drafting the text for new processes, an email announcing promotions or directions around handling organizational changes, or reviews. In this case, you may be prepping slides for a quarterly review to be presented to upper management. If your executive has a social media presence, you may coordinate with your communications manager to schedule when topics and responses are scheduled on various social media platforms. Your executive may be a podcast guest, highlighted in a video review, or publishing a White Paper.

Creating a Career Path: There are many paths for a Chief of Staff. This is only a sample of where a CoS can grow. Some chiefs are ICs (Individual Contributors), and others manage teams. In some organizations, the Chief is a temporary position. Your executive may set the expectation that the role is for a specified amount of time with a path to the next role. Depending on where you are in the organization, your next role could be a general manager or vice president. In some technical teams, the CoS is the predecessor to a COO.

By engaging in learning—online and internally—and having discussions with your mentors on an ongoing basis, you will realize that you have been developing your skills all along.

Sharpening the Saw: What does it mean to "sharpen the saw"? Stephen R. Covey developed this concept in his chapter on Habit 7, which includes a story about sharpening the saw.[4] Suppose you were to come upon someone in the woods working feverishly to saw down a tree. "What are you doing?" you ask.

"Can't you see?" comes the impatient reply. "I'm sawing down this tree."

"You look exhausted! How long have you been at it?"

"Over five hours," he returns, "and I'm beat! This is hard work."

"Well, why don't you take a break for a few minutes and sharpen that saw?" you inquire. "I'm sure it would go a lot faster."

"I don't have time to sharpen the saw," the man says emphatically. "I'm too busy sawing!"

For Covey, sharpening the saw is about renewing and refreshing the four dimensions of our natures—physical, spiritual, mental, and social/emotional—so that we're more effective in our life's work. It's about regularly investing in ourselves so that we can reap dividends on a continual basis. It means working smarter, not harder.

> *"Give me six hours to chop down a tree and I will spend the first four sharpening the axe."*
>
> -Abraham Lincoln

"Sharpening the Saw" can also be called self-care. If you take time out daily and weekly to sharpen your figurative saw, the amount of work you can get done should increase. Taking time daily to meditate, walk your dog, listen to an audiobook, or sit on your favorite recliner to read (whether an industry or non-industry book) gives your mind time to decompress and refocus. I have colleagues who take time to work out midday.

Investing time in what you define as self-care might not look like you are gaining much, but it will benefit you and your ability to

---

[4] Stephen R. Covey: "7 Habits of Highly Successful People"

work in the long run. Some studies also indicate that you won't feel as sick or tired. Besides allowing you to get more done, regularly taking care of yourself increases your sense of agency and effectiveness.

QUICK TIPS

TIP #1: Your RoB should be evergreen and adapted as your business changes. What was needed at inception is not always what is needed in the next phase.

TIP #2: Meetings do not need to be limited to in-person or virtual calls. Define what needs to be discussed and target outcomes—using the various methods at your disposal. An asynchronous conversation, a 1:1 chat, a group chat, or even a well-structured email can provide clarity and outcomes, minimizing meeting fatigue.

TIP #3: Your leadership team should be aligned on what is tracked on the scorecard and the method of dispersal. This provides a unified view for the organization, keying in on the same important metrics.

TIP #4: When taking a meeting break, avoid your cell phone, checking email or anything electronic. Take a walk, exercise, or do something that does not involve using a device or computer.

3

CHIEF OF STAFF PILLARS

There are so many skills you can build to ensure you have a consistent and ongoing career as a chief of staff. Let's look at these pillars, which are not exhaustive but will provide a good glance into what our buffet looks like and why each successful Chief of Staff office will include each of these pillars with some representation. This is one of the fun parts.

Remember the broccoli moment analogy I shared at the beginning? At a buffet, you can choose what you want to dive into. However, before you do, you may have been particular about the type of food you want. Don't be afraid to try something different or new. Your pillars will reflect what is found in your buffet. You will always have more to choose from than you can put on a plate, and similar to the pillars, you will have more work to do than one person or team can execute against.

Your ability to prioritize will help you be successful. These three pillars can be expanded to include the nuances of your role and your teams' expectations.

Our first pillar is Operational Excellence. The team's operation fundamentally runs the business and keeps things flowing smoothly. If things are running well, folks will not notice if you are even doing your job because things are just working as they might expect. Your operational and administrative teams typically are a part of this pillar. You may find your admins, business managers, finance, operations, and, in some cases, vendors.

A general definition is that operational excellence is executing a business strategy more effectively and consistently than competitors. Quite simply, the outcome should be increased revenue, lower operational risk, and lower operating costs. This is an essential strategy to sustain growth while staying ahead of the competition. Regardless of how effective your team is, external factors can block your path forward. These items are activities operated like clockwork and most likely have ties to other cross-functional organizations. Planning is vital and the backbone of a strong office.

There are financial plans to ensure that, from a budgetary perspective, revenue targets are being met. If your team has opportunities to meet with customers and travel, all this must be accounted for and managed.

Another aspect of financial planning concerns vendors, procurement, and hardware and software costs. One chief of staff I worked with was a deputy chief of staff in our controller's office. My team was working on organizing what our financial picture should look like. He provided some amazing perspectives on distinguishing between the budgets we were tracking, our financials, and other aspects of general finance we needed to be aware of. He provided us with feedback on creating guidelines and even templates. Hence, as the team was maturing in their ability to budget for their team, the processes were repeatable, and there were clear explanations. This was extremely helpful.

As a result, either understanding budgeting or embedding a financial director as a part of your team can keep the team aligned with the process and procedures to understand the state of the end-to-end financial picture as you review and understand trends and gaps. This is not the only aspect; when it comes to revenue, there are even more relationships with sales teams, legal teams, and, in many cases, some partner relationships with contracts that need to be tracked and accounted for.

Another final cost that should be factored in is employee costs. Hiring new employees and what those costs translate to in the overall picture will give you a good sense of what it costs to run your business and what needs to be "in the bank" consistently while looking at what may be happening from a socio-economic, political and environmental perspective to keep the business running.

The founder of The Institute for Operational Excellence, Kevin J. Dugan, defines operational excellence as a point where every employee can see the flow of value to the customer and fix it when it is interrupted or breaks down. In an operationally excellent organization, everyone knows their respective contribution to the flow of value to the customer. They can identify when this flow is normal or abnormal and know how to restore any abnormality without the assistance of management. There are many different ways to execute operational excellence, and here are just a few:

Lean Manufacturing: Lean is a project management methodology that focuses on adding value to customers. This is done using value stream mapping, which helps manufacturers determine which product features add value to the customers. Lean manufacturing also uses the "Kaizen principle" of continuous improvement, which involves a cycle of permanent feedback and improvement of any aspect of the manufacturing cycle so that companies are always moving towards operational excellence.

Toyota is a company that recognizes inventory as a significant cost factor for production. As a result, they reduced waste, and excess material was the largest area where they identified waste. By optimizing this excess, they developed lean production. In technology, lean manufacturing principles find application in software development, where efficiency, rapid iterations, and minimizing excess are crucial.

Suppose a tech company adopted lean practices in its software development lifecycle. The company focuses on customer value by emphasizing iterative development cycles, allowing for continuous delivery of features and improvements. Implementing the principles of Lean Software Development, the team reduces "waste" by avoiding unnecessary features, optimizing coding processes, and minimizing delays. Agile methodologies like Scrum promote regular collaboration, adaptability, and responsiveness to changing requirements.

Continuous integration and automated testing ensure the early detection and correction of defects, enhancing overall product quality. By fostering a culture of continuous improvement and empowering cross-functional teams, the tech company can rapidly respond to customer needs, deliver high-quality software, and remain competitive in the dynamic tech landscape.

Six Sigma: Introduced by American engineer Bill Smith while working at Motorola in 1986, Six Sigma is a method that allows organizations to improve their operational efficiency by using the DMAIC process, which stands for Define, Measure, Analyze, Improve, and Control. It is a systematic method that helps resolve operational problems of any kind. Like Lean, Six Sigma also uses value stream mapping and Kaizen as a way to add value to the customer.

Another important aspect of this methodology is Six Sigma 5S, which stands for sort, set in order, shine, standardize, and sustain. This methodology helps organizations improve their business processes. Once again, in the tech space, Six Sigma methodologies are applied to enhance quality and efficiency in various processes.

Consider a software development company employing Six Sigma principles to improve its software testing process. Initially facing challenges of inconsistent testing results and delays, the company conducts a thorough process analysis using Six Sigma's DMAIC framework. The team identifies critical factors affecting testing outcomes by defining clear objectives and key metrics. Measurement tools, such as defect tracking and cycle time analysis, are implemented to quantify existing issues. The company identifies root causes through statistical analysis and implements improvements, such as standardized testing procedures, automated testing tools, and training programs for testers.

With continuous monitoring and control mechanisms in place, the company not only significantly reduces defects but also establishes a more efficient and predictable software testing process, aligning with the rigorous quality standards advocated by Six Sigma.

FLEX Methodology: There's also the flawless execution (also known as PBED, for planning, briefing, execution, and debriefing), which is based on what fighter pilots use in combat. The methodology moved to the private sector in 1998 and has roots in the agile manifesto created by software developers.

As technology evolves, the FLEX approach is finding space in tech. Embracing flexibility as a core principle, a tech company adopts FLEX to enhance its project management processes. Rather than strictly adhering to a fixed plan, the company prioritizes adaptability and responsiveness to changing requirements. FLEX encourages iterative development cycles, allowing the team to regularly reassess and adjust project goals based on customer feedback and emerging priorities. The company leverages FLEX's emphasis on collaboration and communication, fostering a dynamic environment where cross-functional teams collaborate seamlessly, ensuring that development efforts align closely with customer needs.

This approach not only accelerates time-to-market but also enhances the overall quality of the software, as FLEX empowers the team to iterate, learn, and continuously improve throughout the development lifecycle. The origination of the methodology is currently not attributed to a person or organization to this date but is widely used by many industries.

SMART is an acronym for Specific, Measurable, Achievable, Relevant and Timely. Applying this technique helps to have more precise KPIs — measured by actions, not behaviors. They're challenging but attainable, relevant to the organization's objective, and scheduled within a defined timeline. It is generally accepted that the SMART acronym was developed by George T Doran in 1981 in Spokane. He was a consultant and former Director of Corporate Planning for the Washington Water Power Company and published a paper titled "There's a S.M.A.R.T. Way to Write Management's Goals and Objectives."

Early in my career, a mentor helped me rethink crafting my goals using the SMART framework. We took my existing goals and restructured them. Let me share an example using goals and how I was able to edit them. I will highlight two target areas: Better Communication and Leadership Team effectiveness. As you look at the initial goal and revise it, minor adjustments and stronger wording help create, in effect, a "smarter" goal.

Target Goal: Enhance Communication Efficiency	
Initial Goal	*Revised SMART GOAL*
Create a communication plan	*Specific:* Implement a new communication plan to streamline information flow between departments highlighting method of comms, frequency and target audience.
Capture and respond to all internal questions	*Measurable:* Achieve a 20% reduction in response time to internal inquiries within six months.
Conduct regular internal audits	*Achievable:* Conduct a thorough communication audit, identify bottlenecks, and implement solutions.
Identify ways to improve our ability to communicate with the team	*Relevant:* Improving communication aligns with the organization's goal of enhancing operational efficiency.
Implement solutions	*Time-bound:* Complete the implementation and assess results within six months.
Target Goal: Enhance Leadership Team Effectiveness	
Host a leadership retreat	*Specific:* Organize a leadership team retreat focused on team building, strategy aligment and communication.
Increase leadership satisfaction scores	*Measurable*: Increase leadership team satisfaction scores by 15% in the next survey.

Schedule a leadership development workshop	**Achievable**: Design and implement targeted workshops addressing identified team dynamics from employee surveys.
Evaluate the effectiveness of the leadership team	**Relevant:** Strengthening the leadership team contributes to overall organizational effectiveness.
Plan and Schedule a leadership retreat	**Time-bound**: Plan and Schedule a leadership retreat in Q2 and assess satisfaction within three months with 1 key action item.

A key area is strategy planning. While some organizations use Objective Key Results (OKR), others have stuck to Management By Objectives (MBO).

OKRs and MBOs aim to improve organizational performance but differ in their execution. OKRs emphasize setting ambitious, measurable goals and outcomes. MBOs, a more traditional method, focus on cascading objectives down the ladder, emphasizing performance appraisal and employee alignment with organizational goals.

Every organization has goals: what would you like to achieve, and what does that path look like? As a chief, you work as part of a leadership team to ensure conversations are occurring at the right levels with, as required, documentation of your actions, who you plan to partner with, who you need help from, and any metrics or key performance indicators (KPIs). This is where teams will spend time determining how they will see success, even if that success includes some dips in that journey. The leadership teams won't be the only ones driving strategy—they also enlist their teams to define it.

Objectives are a way to break an approach into achievable targets; they are what the organization, team, or individual wants to accomplish and are typically qualitative and time-bound. The Key Results are concrete, specific, and measurable, describing how you'll achieve the objective, the what; and the key results are the how, where, and when.

Offsites and onsites are powerful ways to create a space for targeted conversations. These events have many names, such as leadership summits or retreats, conferences, symposiums, forums, and seminars. In this era of the hybrid work environment, we will call these gatherings "huddles." When I think of huddles, I imagine a team getting together to discuss what needs to be done next—a gathering of key players to determine what is next.

Huddles can be in-person or virtual, depending on the makeup of your teams, and last a few days for geo-diverse teams. This is where you will start to identify which topics the leadership team needs to discuss. Many huddles are multifaceted with the inclusion of development activities, typically to help build or reinforce relationships of trust, operational topics, customer-centric topics, and industry-specific items. For example, a manufacturing or warehouse business may discuss efficiency and quality standards.

❝

Huddles are one part of an extended
conversation — it's about strategies, goals,
and/or tactics. They're a time to bring people
together to make decisions so you can then
go back to the office and start taking action.
If people don't understand (or agree on) the
outcomes they're driving toward or don't feel
their voice is being heard, your huddle will
derail before the first coffee break.

Harvard Business Review states:

The scope of the matters discussed at an offsite strategy
meeting is broader than at the typical management meeting. When looking at big-picture topics like what business
the company should be in and more focused questions like
how to build new core competencies, executives must peer
beyond the immediate horizon to three to ten years into the
future. Instead of concentrating on their functional areas,
participants must take an organization-wide perspective and
synthesize information drawn from disparate areas of the
firm. And unlike operations-oriented meetings, whose objectives are limited and whose function is primarily reportorial
or tactical, strategy offsites deal with information and issues
that are often ambiguous or speculative, which makes many
executives uncomfortable.[5]

---

[5] https://hbr.org/2006/06/off-sites-that-work

ANNE MARIE OTANEZ

Huddles are vital because they allow employees to connect, inspire creativity and innovation, support strategic planning, and strengthen relationships. They help team members see their colleagues in a new light, relax people, put them in a casual mood, and break down hierarchies. Effective huddles provide learning opportunities about employees' passions, skills, and interests. They are one part of an extended conversation—about strategy, goals, and tactics—and a time to bring people together to make decisions so you can return to the office the next day and start taking action. However, if people don't understand (or agree on) what outcomes they're driving toward or don't feel their voice is heard, your huddle will derail before the first coffee break.

I've been able to lead many huddles, and in the age of hybrid, I was able to facilitate one with the leadership team. This was predominantly online, and as a leadership team, we added multiple breaks to help with being on screen all day. Even with relevant topics and high engagement, being virtual can be challenging, and breaking it up with extended pauses in the day can help your team reset and reengage. Later, I will share an outline for a huddle I hosted for my leadership team, which had three goals:

1. Building relationships and connections
2. Alignment of business and strategy
3. A focus on operational excellence

For the relationship-building portion, we enlisted the help of an outside company that spent the day reviewing some previously requested prework that gave insights into each team member and their working styles. We connected during a curated discussion 1:1, in small groups, and with a larger audience. These exercises took the entire workday. As we moved to the next day, focusing on strategy and business alignment and recognizing the need for informal and ad hoc conversations, the agenda was limited to four key topics, filling the day with discussions. This was the desired outcome. Rounding out the huddle was an emphasis on more of the tactical aspects of the day. This hit firmly on the Rhythm of Connections and their associated rhythms of business (covered earlier in the core competencies section).

It is common for big company decisions to be created and driven by executive management. With employee engagement increasingly vital to running a business, embracing your workforce's collective knowledge and experience can prove to be a catalyst for increased performance and a way to keep the team executing at a high rate. Working with staff on strategic initiatives can help them feel valued, improve morale, and make them feel like their opinions are heard, contributing to higher staff engagement and retention. A perceived dictatorial leadership style often leads to employee disengagement and lower productivity. Being in a different environment often helps promote communication between workers and improve professional relationships by allowing people to learn more about the people they work alongside in an external, more relaxed atmosphere.

Huddles boost your organization's morale. They provide a break from the usual work environment and offer a chance to engage in different activities. These activities inspire creative thinking by providing a new environment and breaking the routine of the workplace. They are seen as an investment in the overall culture and provide a designated time and space for addressing important questions or broader goals that often get overlooked in day-to-day operations. So, while huddles might seem like just a fun break from work, they serve several crucial functions that can benefit both the individuals involved and the organization as a whole.

As the CoS, this is a crucial event that you will lead planning efforts in collaboration with the leadership team and key stakeholders. You can survey the team for ideas or work with your executive if a theme or set of topics needs to be discussed. You will want to ensure a goal(s) and outcomes for the huddle and the individuals who need to attend.

Some huddles will require some prework, so allow ample time for the team to complete the prework without being rushed. Some huddles can be one day to a few days. Breaks are essential, and including some non-curated time will help with some informal discussions. Since your team may be traveling, you will want to support helping them maximize their time.

This agenda should identify a mix of strategic and operational topics that should be reviewed with the leadership team regularly, whether annually or biannually, with a follow-up plan.

Rounding out this pillar highlights the need to ensure that compliance is adhered to. There are many aspects from a security, regulatory, and legal perspective that this area gets attention. For example, Technology companies operating in the EU must ensure they are GDPR compliant. Whether your team or a cross-functional team handles this, any regulatory rule must be followed. Depending on your organization, you will want to familiarize yourself with what is required to keep you compliant and informed.

Compliance is crucial in business for several reasons. Compliance ensures that businesses operate within the boundaries of legality and ethics, safeguarding the interests of various stakeholders. It involves adhering to local, regional, and international laws governing business operations. It also helps to identify and avoid possible red flags in your business. Non-compliance can lead to legal repercussions, including fines, penalties, and lawsuits. It also assists with business or product reputation. This promotes transparency, accountability, and fairness in business practices, enhancing the reputation and credibility of an organization. Non-compliance can damage the organization's reputation, erode customer trust, and result in the loss of business opportunities.

Companies must comply with relevant guidelines to establish trust with customers, investors, employees, and regulatory authorities. Corporate compliance also helps your employees act responsibly. The best corporate compliance program teaches employees to treat each other well at work, promote high professionalism, and uphold corporate values inside and outside their workplace.

Compliance is critical for the health, safety, and well-being of employees, customers, and the general public. It ensures that the organization adheres to internal and external rules and regulations. Compliance is required for public services to be delivered efficiently and effectively.

The next pillar is around Communications. This pillar is a great entry point for conversations, listening rhythms, and anything related to messaging around your organization, whether it's for internally or externally facing individuals. Depending on what your organization is doing, all your communications may only be structured around internal communications. They can be a little more informal so that anyone can consume them.

You want to understand your organization's listening rhythms to keep things running smoothly. Many teams have either all-hands meetings or organization-wide meetings to discuss the state of the business, highlight new individuals in the organization, and highlight organizational updates related to personal things that are happening with individuals, e.g., newborn babies. Though called different names, both meetings cover the same types of information and are mainly a medium to disperse information to a large group of people.

This is also a great time to highlight wins, and if there are not many wins but learning opportunities, this is a pivotal time to discuss it as an organization. In one all-hand meeting, the team focused on getting a specific customer. The org-wide meeting was a celebratory discussion highlighting this desirable win for the team and the many people who contributed to the success within and outside the team. Because these meetings are in an auditorium and online, there was audible cheering from the team. Wins can be very unifying.

As the chief of staff, crucial to your role is helping prepare this content and, on some occasions, soliciting guest speakers or panels to talk to your team to provide cross-functional perspectives on their interaction with your team and that strategy again, which is vital to the success of your business. Some organizations have coffee talks or AMAs, which are "ask me anything" sessions.

Those tend to be very informal, but also an opportunity for individuals to ask questions with a leadership body where the content has not been curated. Depending on the ability of people in your team to feel vulnerable, some of the questions will be very frank, and those are the best meetings because they show a level of psychological and emotional safety that makes people feel okay asking those hard questions. There are also special fireside chats that engage certain groups (i.e., all managers to discuss specific information). Your goal is to ensure that there are multiple listening rhythms for any staff member to access a variety of executives and feel a sense of connection to the vision, mission, and leadership body. These are just the audible ways to ensure listening rhythms are in place.

There's also a variety of newsletters that highlight things that are happening that may be curated. You may want to guarantee that there is a library where these newsletters are contained so an employee can access them virtually at any time. Sometimes, some blogs are accessible to individuals, but again, you want multiple entry points and ways for people to interact when face-to-face is not always an option.

Another aspect of communications is the items your team needs to be privy to. I delivered a monthly newsletter to the leadership team and key managers. This newsletter would often contain, in one place, information about new guidelines, processes, and procedures, as well as links to white papers and documents that refer to policy updates or changes, vital upcoming meetings and events for that month, and the subsequent months as a heads up.

Depending on your organization, you may publish business scorecards, KPIs, OKRs updates, and any planning or strategic discussions requiring prep time. Budget and financial details with trending information are also helpful for the team to ensure alignment—calendar information around office closedown, team events, offsites, and retreats.

Change management is an essential aspect of this role. Significant changes often occur, such as reorganization, growth due to scaling up or size, creating a significant product/program, etc. You should employ change management for anything that will significantly impact your teams and your organization. The good thing about a change management plan is that it is effective regardless of impact size. A change management plan is a plan that outlines how you're going to control changes to the organization as a whole or the scope, goals, activities, budget, and resources involved in a project.

A change management plan contains:

- Description of the change
- Background and context for why the change is being adopted
- Possible areas of impact
- People and teams involved
- Budgets and timeline for the change process
- Anticipated goals and outcome

A Change Management Plan Template is a pre-designed document that provides a structured format for creating a Change Management Plan. Change management is the overarching plan for coordinating budget, schedule, communication, and resources.

Another method of communication is a brown bag meeting. It is referred to as a brown bag meeting (or a brown bag seminar) because participants typically bring their lunches, which are associated with being packed in brown paper bags. Brown bag meetings are typically informal training and learning sessions. They don't necessarily have to occur during the lunch hour; they can happen at any time during the workday or after hours. Brown bag meetings promote dialogue and information sharing among participants and usually are held in conference rooms. Sharing among participants enhances training and ensures consistent information spreading while encouraging teamwork, reinforcing company values and missions, and increasing employee morale.

Cultural growth and diversity seem to have taken root lately. Many organizations will identify individuals or departments that handle both opportunity areas. As the chief of staff, you can play a critical role in building and maintaining an inclusive culture that recognizes the diversity that people bring. The one rule to remember is that it is not the job of someone from a marginalized community to lead these efforts.

Let me say that again: Just because you are the Black person in the org does not mean you lead the "Black community initiatives." Just because you are the neurodivergent individual does not mean you lead the "neuro-divergent initiatives." Or just because you are the only non-binary individual does not mean you lead the "non-binary initiatives." As a member of these communities, not finding yourself in a situation where you are a driver can be a tricky balance.

Leaders face several challenges in promoting diversity and inclusion within organizations. Some of these challenges include:

General resistance to a change: Implementing diversity and inclusion initiatives often requires a shift in organizational culture and practices. Resistance to change from employees who may be accustomed to existing norms can be a significant challenge for leaders. We talk a lot about biases. Leaders may unintentionally harbor biases that affect decision-making processes, such as hiring, promotions, and project assignments. Overcoming unconscious bias and fostering unbiased practices is a continual challenge.

A lack of diversity at all levels of the organization, especially in leadership positions, is a common challenge. Limited representation of diverse groups in top leadership can hinder the development of inclusive policies and practices. The absence of effective diversity and inclusion policies or the presence of policies that are not actively enforced can impede progress. Leaders must ensure that policies are comprehensive, communicated effectively, and consistently implemented. Leaders must be cautious about avoiding tokenism, where individuals from underrepresented groups are included merely to give the appearance of diversity. True inclusivity involves creating an environment where everyone's contributions are valued.

The measuring process can be tricky. Quantifying the impact and progress of diversity and inclusion initiatives can be challenging for an organization. Leaders need to develop effective metrics and regularly assess the success of their efforts, but they should not be doing surface or strictly performative work.

Retaining diverse talent and ensuring equal opportunities for career advancement are ongoing challenges. Leaders must address barriers preventing certain groups from thriving within the organization. Some employees may lack awareness or understanding of the importance of diversity and inclusion. Leaders need to invest in education and awareness programs to foster a more inclusive workplace culture.

Overcoming these challenges requires commitment, continuous effort, and a comprehensive strategy that involves leadership commitment, employee engagement, and a commitment to fostering an inclusive workplace culture.

"

Marginalized people often find themselves at the center of explaining, teaching, and providing assistance for these national and globally traumatic events. Having to be in a role responsible for supplying ongoing education leads to what is known as 'vulnerability' fatigue.

As executives look to find out how to make sure these voices are heard, one option is for some companies to look for a person in the community to lead. As a person from a marginalized community, I choose to lead and drive these initiatives. Great perspectives are found in each person, but be mindful of how much that person is called on to lead, direct, or share. From the marginalized person, leading and driving these perspectives can lead to what is known as "vulnerability" fatigue.

This type of fatigue opens the door for discussing feelings that are not directly related to the event but have roots in some element that a person can relate to. Each time an event occurs, you are drained. And regardless of the day, depending on your financial or familial situation, you may have to work.

For example, in the year following the death of George Floyd, 229 Black people lost their lives to police officers across the United States[6]—a number which has only increased in the past three years, including the most recent murder of Sonya Massey in the month prior to the release of this publication. So, to provide a glimpse of what this level of fatigue could look like, let me share some thoughts I penned on LinkedIn following one of these events:

Breonna Taylor, a 26-year-old Black American woman, was fatally shot in her Louisville, Kentucky apartment on March 13, 2020, when at least seven police officers forced entry into the apartment as part of an investigation into drug dealing operations.

It is 9 o'clock, and I have already been on a conference call this morning. I take a quick commute to my kitchen (my new normal) to grab a cup of tea and a slice of toast before joining an all-day call with leaders providing updates to the Sr. Executive in my organization. I am excited because it is the first of many of these calls I can participate in, and it gives me a glimpse into what the 10,000-foot view looks like. I have spent many years "in the trenches," which offers a unique vantage for me at this stage in my career. One of my team members is presenting with a large group, collaborating on several features later in the afternoon. I make

---

[6] www.newsweek.com/full-list-229-black-people-killed-police-since-george-floyds-murder-1594477

sure I am set up to catch up on some work while I listen. Around noon a friend sends me a text "Thinking of you today, and I just want to hug you and be pissed off with you." I didn't even have to know the context of the text; I knew there would be an announcement of the Breonna Taylor verdict this week. I had been so entrenched in my workday that I was not distracted by any news that day. I had deleted several social media accounts in the previous years, so if I wanted to know what was happening, I sought out what I felt were reputable sources.

I took a breath and opened my browser. I had to take a beat. I read a few more articles and sat in disbelief in front of my monitor. I had thought it might be different this time. I had hoped it might be different. The movement was not only made up of Black people. There was and is a growing group of allies also enraged. The numbers were growing. This is STILL being talked about every day. People are, daily, STILL getting activated and trying to do what they can and DOING what they can. Black lives have stopped trending. We knew this would happen, but the killings are STILL happening at a high and alarming rate with little to no respite.

It was still about Breonna, and the news clip I read said, "On March 13, 2020, Breonna Taylor, a 26-year-old EMT, was killed by the Louisville Metro Police Department. She was asleep. There have been protests every single day in Kentucky for justice for Breonna. She is one of many killed by the police department."

Yesterday, after months, the response: *As reported by the Washington Post on September 24, 2020 (edited)*

> The grand jury in Jefferson County, Ky., indicted Brett Hankison, who was fired by the department in June, with a termination letter saying he "wantonly and blindly" shot ten times into Taylor's apartment. He is accused of endangering lives in a neighboring unit by firing the rounds.
>
> The other officers involved in Taylor's death, including the one determined to have fired the fatal shot, were not indicted. Kentucky Attorney General Daniel Cameron (R) said he did not anticipate charges in the future.
>
> Cameron said the state's investigation determined that the officers who shot Taylor were "justified" because they had been fired upon first by Kenneth Walker, Taylor's boyfriend. Walker has sued Louisville police and disputed their version of events.

I stopped reading the articles online and realized I had missed the one presentation I wanted to hear and couldn't breathe. I literally feel like I cannot breathe. I look at my monitors; one is flashing the meeting details, and the other is my Ring cameras. Yes, I have cameras that point to every entrance of my home and a view of the backyard. I look at this daily to see anyone approaching my home. I never feel safe and am only appeased when my husband is home.

I sent a note to my manager expressing some of my feelings, and he, fully supportive, shared my angst. I walk around my home, and my eyes fill with water.

That young girl was home hurting no one and woke up to her maker. She was a Black woman in her sanctuary, which was not a sanctuary for her. My heart hurts because I am her. I look at the Black women in my life; we are Breonna.

By the way, that day-long workshop is still going on, and I have missed most of it. Yes, it's recorded, but how often do we take time to listen to the recordings?

In case I failed to note, there is a pandemic going on . . . and even with all the attempts to ease the ever-growing congestion of meetings colliding with focus time, I still need to update documents, review this PowerPoint deck, and respond to emails. I try to cram twelve hours into a nine-to-ten-hour day. We all do this and try to find the time. We all find the time. The reality of the situation is that we all have some intersections in our lived experiences.

My lived experience, while it affords me the opportunity to take a step back, reminds me that I have a mortgage and family to take care of. It reminds me that I have parents that I choose to help. I can donate money to organizations that are at the forefront of fighting and exposing the racism that has been sewn into the fabric of this culture. I have conversations with Black peers who struggle to find the words. I have conversations with non-Black people who struggle to understand and work on empathizing. We offer each other encouragement and share resources to deal with, sometimes, just that moment. I am also reminded that I am not alone in this fight.

While I may not see the change in the world I want for me, I will fight to see the change I want to see in the world for my nieces and their cousins. I need to stay undaunted.

Yesterday is now gone, and today, I have choices I need to make. I need to remain undaunted in my efforts and choices. We all have choices we need to make. Today, I have more meetings and work that needs to be done. The "day" is the great equalizer. It does not stop when things go wrong, and it does not stop when things go well. It cranks along, and we either let it run us over or take it by the horns and try to lead it down a path.

I am reminded of the words of a great woman I was lucky to see speak in person, Maya Angelou: *"And still I rise!"* I will continue to take the day and lead it down a path. Some days, I am more successful than others, but I remain undaunted.

Is there an opportunity to just stop working? Can I take a break? Depending on your organization, there is not usually space to take a mental health day or week to digest what has happened. Now imagine this happening daily, weekly, or monthly.

As I wrestle with the daily things that will challenge my day, I am reminded to care for myself. I will have moments where I feel overwhelmed, and that is okay. This is one dimension of me during the workday. I internally work on the relationships I have at work so when I have moments like this, I can take a step back, whatever that might look like, regroup, and if the organization or team does not lend itself to that, I connect with mentors and sponsors within my network who will give me the space to be just me.

How does a leader thrive in this environment? There are many available resources to help. As a CoS and if you come from a marginalized community, those individuals who are allies and sit in those positions of power should do the heavy lifting to ensure that the changes that are occurring set you up for success and you can drive where it makes sense and not where they simply need a face.

Learning and education are available by a variety of DEI leaders who can provide an objective view and infuse what is needed to not only make a change but also reinforce the change and not be something fleeting or trending but have a lasting and enduring impact. This will empower that employee to be in those spaces that do not necessarily retraumatize them but allow them to be their authentic selves in a space that they feel comfortable in.

There is so much that can be said about culture and community. A team's culture includes not only the things that are seen but those that are not seen *or* heard. Some examples of ways to understand and navigate the cultural variables in your team:

1. Consider distributing documents to the team for review before a meeting. Pre-reads provide space for those who can provide excellent feedback but need time to absorb the content. I worked with a person who needed additional time to consume information rather than being in a meeting with a limited and finite amount of time.

2. Solicit feedback in advance from those who may not naturally speak up. For hybrid meetings, elicit the icon highlighting those who want to comment (i.e., in a communication platform, some communities have a "raise hand" feature, prioritizing based on when the hands were raised).

3. Setting an agenda and asking the best way to exchange ideas. You may have a team that is effective asynchronously and can interact and bring out the best in each other rather than sit in or dial into a meeting.

These are just a few examples. Ask your team questions and make it a priority to understand the best way to communicate is an investment in the community you are creating. Building a community within a team involves fostering a sense of belonging, trust, and cooperation among team members.

"

When culture meets community, there is harmony... and ignorance loses its power as inclusivity becomes the norm.

Also, share praise often, publicly and privately. When you understand the work behind what your team is doing, praise becomes easier because the sense of appreciation will come naturally. "Thank you" and "I really appreciate..." will become regular phrases in your dialogue. People who feel appreciated and rewarded have a sense of motivation and will enjoy what they do because they are connected to the strategy and vision and genuinely feel a part of the team and organization. As you keep up with this focus, the community will grow.

Culturally, as you embrace where your team is from and what they represent, ensuring they have a voice, your work will represent the customers you serve because what they see in your organization is demonstrated in the work. When culture meets community, there is harmony, and ignorance loses power as inclusivity becomes the norm.

ANNE MARIE OTANEZ

Org strategy is the most important pillar, but it has nowhere to go without operational mechanisms and solid communications vehicles in place. This is where the "why" meets the "how." Why are we building what we are building, and what does it all matter? You are creating initiatives to help you secure market advantage, build a revenue line, increase your user base, or achieve whatever target you hope to secure. As the CoS, you want to ensure the benefits and rewards are clear.

Every business leader wants their organization to succeed. Turning a profit and satisfying stakeholders are worthy objectives but aren't feasible without an effective business strategy, which is crucial for success. Here are some benefits of having a business strategy:

Value: A business strategy helps create value for the organization, its customers, suppliers, and employees. It's built around three key questions: How can my business create value for customers? How can my business create value for employees? How can my business create value by collaborating with suppliers?

Advantage: A well-crafted business strategy gives companies a competitive edge. It helps set organizational goals and determines various business factors, including pricing, sourcing materials, employee recruitment, and resource allocation.

Strategic Planning: Strategic planning is an ongoing process that uses available knowledge to document a business's intended direction. It helps prioritize efforts, effectively allocate resources, align shareholders and employees on the organization's goals, and ensures those goals are backed by data and sound reasoning.

Alignment of Stakeholders: One significant benefit of strategic planning is creating a single, forward-focused vision that can align your company and its shareholders.

Operational Efficiency: Strategic planning helps to increase operational efficiency.

Market Share Growth: A sound business strategy can help grow your market share.

Profitability: Strategic planning can increase the profitability of your business.

Remember, a company without a clear business strategy is unlikely to succeed. Therefore, a thorough understanding of value creation and paying proper attention to strategic planning is necessary.

Many elements make up a good business strategy. Identify and dedicate yourself to a plan. Before you get into the details, work with your team and key stakeholders to collaborate with the cross-functional team and ensure that every decision-maker is aligned on the strategy and plan. The benefit of having this in place is a shared vision. You will want to review your workforce plan as you work with executives to create this shared vision. Who is working on what and which roles will warrant a high success rate and that metrics are met and trend positively? If not, you can pivot quickly. This should line up with your strategy.

Formulate those goals and associated metrics. Most teams will define this for the upcoming fiscal year. Defining the plan for 3-5 years, revisiting it, and adjusting it is vital to gaining success. Don't get stuck on the same goals year after year. Economic, financial, and many other factors can impact how you create and keep momentum.

In 2020, the COVID-19 pandemic caused every business to adjust how they did business. Some survived, and for other companies, it caused their doors to close. Your ability to adapt quickly to changing conditions can safeguard your business's stability.

We discussed business strategies, but there are also functional strategies. Each team will have detailed plans that ladder into the business strategies. They highlight the functional areas and teams needed to execute against the overarching strategy. You will use data and information to understand which strategies are the strongest. Using analytics will form some of your opinions, and a solid change management approach is essential because as your organization changes and adapts, you can predict the adoption of whatever change is necessary.

In essence, change management strategies define the approach needed to manage change, given the unique situation of your roadmap. Implementing new processes, products, and business strategies is critical while minimizing adverse outcomes.

A roadmap is a plan that outlines a plan or strategy for achieving business goals. It is a visual representation of a strategy that answers questions such as what will be done, who will be involved in the work, the scope and resource allocation details, and how and why certain initiatives were prioritized over others. Roadmaps can help smooth alignment, improve strategic organization, and centralize team collaboration—regardless of what kind of business you have.

Primarily used for planning projects and developing new products, roadmaps are created and presented to align all stakeholders and your entire team on one strategy. The basic definition of a roadmap is simple: it's a visual way to communicate a plan or strategy quickly. Every team has a plan and strategy built around doing what pushes the business goals forward. You can get lost in day-to-day task management when you're busy executing the strategy. A roadmap is one of the most effective tools for rising above the granular details and chaos. Roadmaps give you a bird's-eye view of everything happening at your team or company.

A good roadmap should effectively communicate alignment, resources, estimates, and dependencies with other teams.

CSAT: Customer satisfaction surveys are essential for keeping customers on your side by providing valuable insight into how customers perceive you. They help you understand what is working across your business and what isn't, allowing you to improve on the negative and double down on the positive. By opening the door for your customers to have their say, you'll get ideas about other avenues to improve your business—such as new products or services. When you invite feedback, you'll learn what makes your customers satisfied and, more importantly, what does not satisfy them.

Effective customer surveys will help you identify opportunities to improve specific points in the customer journey. They also tell you how your business compares to your competitors. They give you a chance to develop brand loyalty.

Customer engagement is the process of building relationships with customers at every touch point. It involves understanding customer needs, preferences, and pain points. It's about tailoring brand-related experiences to meet and exceed those expectations. The engagement process begins when a potential customer becomes aware of a brand or product—and extends beyond the post-purchase stage. Not only does it help with acquiring new customers, but it is also meant to help nurture and retain existing ones. The ultimate goal of customer engagement is to build lasting and mutually beneficial relationships with customers, leading to an increased lifetime value for the brand.

Customer planning, often referred to as a customer success plan, is a strategy that lays out what customer success looks like and how you'll help your customers achieve that success. It's a roadmap that customer success teams use to operate, ensuring that the right resources are being delivered to your customers so that they will receive value from using your product.

The customer success plan is different from customer service or support. While customer service is a reactive function that handles unanticipated issues or needs from customers, customer success is a proactive approach to making your customers successful. Creating a customer success plan can help reduce churn, improve opportunities for upsell and cross-sell, and increase recurring revenue.

Businesses often face several challenges when it comes to customer planning. It is a significant challenge to know what customers want, what they do, and why they do it. Excellent customer experience always starts with the customer. Identifying the basics of a customer experience strategy is crucial, but it can be challenging. Where do the opportunities lie, and where are the challenges? Ensuring all employees are well-trained and knowledgeable about the company's products or services is essential for excellent customer service.

For one of the teams I worked on, I needed to understand our business strategy better than I already did. I was fortunate to have a leader who wanted me to be on this journey with him. We read through our strategy and vision documents together, and I could ask him the dreaded stupid questions. I then took some time just to let the information sink in. Subsequently, I held a training with my direct reports to discuss strategy. My leader attended the meeting, and I asked him to fill in the blanks where I may not have the full context. I then was in a position where I was teaching my team our strategy based on the understanding I had gained from conversing with my leader. It was a great experience because I had to answer a few questions, which made me delve into what I understood.

Once you've learned it, the best way to understand a concept is to teach it, and I was in a position where I had to teach what I had learned. Now, to be honest, he did have to step in a few times to help fill in some of those blanks, but afterward, we discussed it, and he was thrilled at my grasp of the basic concepts and understanding of our business. I understood our customers and how we were answering their needs.

Often, different departments within a company operate independently of each other, which can lead to a disjointed customer experience. Today, customers have a nearly limitless array of services and products. With this abundance of choices, consumers expect companies to deliver a high level of service consistently. Overcoming these challenges requires strategic planning and execution.

Workstreams: The Oxford Dictionary defines a workstream as "a particular project, process, or area of operations within a business or organization." Workstreams are a way to:

- Divide and conquer large projects.
- Cut your project into slices you can handle more easily.

The "work" in the workstream refers to the work your team needs to do on its piece of the project in order to finish it at the desired level of quality. The idea is that each team handles its own project piece and then passes its part on to the next team in line.

This makes it much easier to manage tasks than if they were all done by one group since each group only has to be aware of its part of the overall project, allowing you to split the work into segments so that teams can be assigned based on their expertise.

QUICK TIPS

TIP #1: You should always be in learning mode. Balance job duties with the skills required to support your executive, identify gaps, and target those skills; as a chief, you need to continuously improve.

TIP #2: While the strategy is being defined, identify which framework you will use to track whether your team is hitting its desired outcomes and where there are gaps or room for improvement.

TIP #3: Define the pillars in your organization and their segments. This will provide you with a list of top areas for work and maturity, as well as a method to scale for growth.

4

VULNERABILITY FATIGUE

W hen there is an event that is being showcased—acts of violence against Black or Brown people—it is more than likely featured in the news, in podcasts, and trending as a new hashtag. There will be a video that is shared more times than it should. If the person in the video has had an act of violence committed against them, they will also have their past and present highlighted for all to read, often irrespective of their part (most of the time, they are the victim). Their family also has their privacy invaded, and pictures are all over any outlet that will share them. And then what?

As a Black woman, I am often the one tasked with sharing the story, the journey, and the hurt. I have put myself out as an employee and marginalized person reacting to the situation while many things are occurring. Sometimes, a person you may not have a relationship with will want to offer sympathy. This person will ask what they can do and how they can help. This is not bad, but it puts you in the position of education and opening up yourself to explain what is occurring in the community.

This is an important point whether you come from a community that feels the burden of shouldering the work of representing your community or you are an ally. Many experts can discuss these topics in-depth and provide not only socio-economic or clinical perspectives but also leaders who offer in-depth views on how to recognize and combat the issues that keep marginalized people traumatized and limited in their access to resources and wealth.

When I look at vulnerability fatigue, every person from a marginalized community has felt this in some capacity. Whether from the viewpoint of "I need to raise the banner for awareness" or "This event did just not occur," now a well-meaning coworker wants to check in or create their narrative and needs validation.

Do we really need to wait for something terrible to happen? Are we not in the business of creating community and elevating culture today? If the tragedy is the catalyst, then you are behind. If your team builds products for consumers, then your team should reflect that consumer base. If not, you are missing the mark. When your team reflects on your customers, you are starting to create a space for diverse voices and views. In that creation, there will be disagreements. Those disagreements do not mean negative conflict but healthy conflict in the team.

Conflict is healthy when it aims to improve the team's outcomes, is respectful, and is not personal. It's healthy when it's out in the open, visible to all team members, and available equally so everyone can safely participate. Healthy conflict requires openness and an ability to entertain others' ideas. When there is healthy conflict in the workplace, teams experience more productivity and engagement thanks to more innovation, creativity, and collaboration.

Conversely, negative conflict is destructive, unhealthy, and unproductive. It occurs when the parties involved have different intentions and goals and use harmful methods such as threats, verbal abuse, and deception to achieve them. Negative conflict can lead to bullying, harassment, discrimination, reduced cooperation, low morale, and poor self-concepts.

Teams that are diverse in regard to not only the visible differences but also those differences that are not readily observable represent a broader base. Those teams with perspectives that, although they may not see eye to eye, end up being complimentary because they bring out viewpoints that may be missed in a homogenous group.

Early on, we discussed the chief of staff as a conductor. Violin quartets are beautiful, but how much more remarkable is the sound when a brass, woodwinds, and percussion section is included? It is not to take away from the magnificence of the violins, but you get a different sound. When a choral group is added, the sound has a certain fullness in that orchestral majesty. All who listen will key into some aspect of the sound that was particularly moving while appreciating the fullness of the piece. We embody and reflect what is being sent to us.

This concept is the same with a team. Have you ever bought a product and, after using it, asked yourself, "Who decided on this?" A simplistic example to emphasize this point is the input of a group of men who could have good perspectives when designing a product for women. However, including women in the equation will allow for an important voice and viewpoint that may not be shared. Thus, if your leadership team makes sure every voice is heard at the table, it must represent the community it serves.

I believe that every voice is not just a token but a valued team member with useful and implementable insights. If there is actual representation at all levels of the business, then when these moments of tragedy occur, there is no lift required to show compassion and empathy. The team has been and continues to build that muscle, and the ask of "Are you okay?" is felt and seen as genuine, and that marginalized person then knows they have the space to be authentic.

Much performative work done by people and organizations is genuinely harmful in the name of partnership. Performative allyship is the practice of words, posts, and gestures that promote an individual's virtuous moral compass more than actually help the causes they intend to showcase. It is very self-serving and inward-focused instead of outward-focused. It is a term used to describe when someone professes to support marginalized groups for reasons other than genuinely caring about their experiences and equality. Performative allyship can be harmful because it can lead to tokenism, exploitation, and a false sense of progress.

> Understand that unless you have a very specific and exclusive business model, or your target is limited, your audience will always be diverse. It doesn't matter the service, the product, or offering.

I remember reading a report about increased Black representation in a specific field. Leadership boasted a 50% growth rate. While that sounds impressive, a 100% growth rate of 1 is 2, and in an organization with 4000 people, is the percentage the number that should be boasted, or should we recognize that there is still work to be done? Details and context paint a fuller story. Do you mandate diversity training, or are employees able to opt in? Which option opens the door for performative connections vs genuine empathy?

When the actions are performative, then there is re-traumatization. Those marginalized people know when they are the token and when someone is professing without action. I did. It never felt good, and regardless of my job situation, it highlighted the divide and the lack of equality. I found myself practicing equanimity in the face of harm by those who thought they were doing good. These are the same individuals who will let you know they are your ally and remind you of this. We don't need to be reminded.

This is a mentally and physically draining place to be. Why? Unlike the other skills you are developing as chief of staff, this draws on those parts of you that are personal, important, and sometimes the intimate parts of your persona. Those are the parts of you that make you who you are regardless of title, income, or familial standing. This taps into the essence of your being.

The fatigue, in this sense, is very exhausting and can feel personal. This is now pulling out what is beneath it all to help provide space for someone who doesn't have your lived experience and perspective. If someone criticizes your viewpoint on running a project, it may be personal, but it's about a skill you naturally cultivated or learned. If a person criticizes your stance on a cultural issue, it feels personal.

Sometimes, insensitive things are said to help you "get over it." Statements like, "It's not personal," "You are being too sensitive," "It's not just Black people," and many, many more statements like these that could fill a book end up feeling like a weight around your neck as you look to provide a perspective. Enter vulnerability and do this enough... *fatigue*. Sometimes, to shift the attention away from the situation, the professed ally will transfer the focus to them.

Enter a sigh! What do we do about this, and how does one combat vulnerability fatigue in a team? Most of the leaders I have supported have been white men. And to share a way that I have tried to combat vulnerability fatigue, I was conversing with one of the leaders of a team. He, too, was a white gentleman and had worked at the company for many years. We had mainly talked about work-related topics and very infrequently touched on our personal lives.

One day, in one of our connections, he mentioned he was struggling to make sure he could find diverse candidates to put in the interview loop. We talked about where he was sourcing his candidates and the assistance he was getting from HR. Our conversation wasn't long, but he said he would continue looking.

Following this interaction, I had a discussion with a leader we both reported to and said, quite frankly, "I can't have this conversation with every white man who doesn't know how to source or find communities of marginalized people to find exceptional talent. You need to have this conversation with him." My executive listened and said, "Absolutely, and I will take that on." This may seem like it was not a big deal, but it was. As I stated earlier, it frequently feels like a marginalized person's job or duty to help non-marginalized people on their journey. At the end of the day, it is everybody's job to figure out how to ensure that they are building a diverse and inclusive space, not just the people who are not being seen in those environments.

Start at the beginning. If you are the leader of an organization, put measures in place to hire, retain, and promote diverse talent not only outside the organization but also inside it. All movement does not need to be lateral. Build the community from the start and not when there is an incident. Identifying the communities represented and acknowledging holidays and festivities despite your nationality is a step toward inclusion. As a chief, you can take the lead in this arena. What are the discussions being held, who is included, and more importantly, who is not included and why? Be judicious!

Understand that despite your business model, unless you have a very specific and exclusive model (which is fine) and your target is limited, your audience will always be very diverse. The service, the product, or the offering doesn't matter. If you are not already thinking about this, then start now.

Let me say it again: *regardless of the service, product, or offering, your audience will be diverse.* We live in a time of many work types: office, hybrid, and fully remote. Each type has pros and cons, and some roles require a physical presence for the work being performed. Challenge the way for the fully in-office roles. Is it about creating and maintaining a connection or micro-management? As discussed earlier, you will build rhythms of connections for your organization. Look at opportunities to "huddle" with your team or key leaders at specific times to develop camaraderie and connection.

When you build onboarding plans, if you are hired regardless of location, there may be opportunities to bring in the teams at different times to foster team building. Be creative. Hire across the globe and build connection points for face time. The book on effective connections in a hybrid world is still being written. Bring your innovative mind and include your organization for ideas.

QUICK TIPS

TIP #1: Build a culture of community at the start so that when challenges arise, you have already practiced and are exercising compassion and empathy.

TIP #2: If you're a marginalized person leading the charge, recognize you don't have to. Balance your time and perspectives. Everyone should lean in.

TIP #3: Identify what your community looks like today and celebrate the differences.

"

Good communication and leadership are all about connecting. If you can connect with others at every level... your relationships are stronger, your sense of community improves, your ability to create teamwork increases, your influence increases, and your productivity skyrockets.

-JOHN MAXWELL,
AUTHOR AND SPEAKER

5

KEY PARTNERSHIPS

~eee~

Many chiefs work for a specific executive and sometimes a team of executives. This is your primary connection. A chief's uniqueness is the connection's success with your executive, which can make or break this job. You are the eyes and ears of this executive. Your job is to:

- Anticipate, act, or recommend a course of action.

- Understand your team, your org, and your executive, so that you will see and make the links to influence.

- Recognize who your executive's important contacts are and how they spend their time.

- Determine who your executives' delegates are, and if those have not been determined, review the organizational chart, provide some assumptions based on team charters and work, and identify the delegates.

Your executive will want to be involved in everything but also know which discussions are critical and which are not. As you understand strategies, you will want to build your network with other key partners with whom your executive works. For example, if your executive is a senior director and collaborates with other senior directors, find if those executives have a chief of staff and work on the connection. Follow this pattern with all your executives' key relationships.

Your leadership team will be made up of your peers. Other key partners will be groups that may support the work your team is engaged in, but these individuals are matrixed. You will work with Human Resources, your recruitment team, the legal team, and countless others who will help you run your organization.

As you work on the cross-collaboration efforts with these partners, you also establish connections with these key folks and your team members. I have found it helpful to ensure at least a monthly cadence where partners and critical executives connect and discuss projects, initiatives, and other important topics associated with the business. This will be an aspect of the company that may see change. Depending on what is happening within the organization, your key partners will play different roles in supporting the initiatives and connecting with you to achieve the vision and strategy.

Typically, your partnerships are teams and leaders who do not have a direct reporting relationship to the leadership team. The number of teams you collaborate with is not limited to team or size. The most crucial aspect is impact. How are they interacting with you and the business model? They have a vested interest in the success of your business. Look for opportunities to include them in pivotal discussions, huddles, and other crucial considerations.

Take the time to understand their organizations and what they need from your team to provide the necessary support. If you are working with a compliance team, understand which checks need to be run, when and if compliance is not met, and the consequences. It is never recommended to be out of compliance, but you assess that certain risks will be considered. Ensure that you understand the complete picture and can communicate all risks so informed choices can be made.

If you work with the marketing department, there may be conferences that your executive or team will need to attend or even be a presenter for. Understanding when these events occur and what is required will necessitate that there is sufficient prep time in advance. Headshots may be needed in advance, and biographies may need to be written.

Working with human resources, as you assess your organization's health and growth, this picture will help you work with the leadership for organizational changes, hiring and retention activities, onboarding, potential job reductions, and growth into geographical areas that your team may not be in already. This will also help you assess where team responsibilities may lie and where they are needed. As you detail these working relationships, create time for these partners and specific individuals on the team to have an ongoing connection.

Some organizations will authorize governance boards to facilitate these conversations. A governance board is a group responsible for overseeing and directing a team. It contains systems and processes that govern the behavior, scope, and decision-making of the organization they represent. While governance boards can be highly effective in driving the organization, centralizing authority can be problematic. It can lead to the concentration of decision-making power in the hands of a few individuals or groups, which can result in the marginalization of other stakeholders and the exclusion of their voices from the decision-making process.

The chief of staff plays a pivotal role in governance boards, acting as a strategic partner and facilitator between the executive leadership and board members. With a set of responsibilities, the CoS contributes to the smooth functioning of the board by managing communication, facilitating decision-making processes, and ensuring alignment between organizational goals and board directives.

This role requires a unique blend of leadership, communication, and organizational skills, as the CoS is a trusted advisor to the executive team and the board. By fostering collaboration, streamlining workflows, and providing crucial support to the governance structure, the chief contributes significantly to the effectiveness and efficiency of governance boards in navigating complex organizational landscapes.

QUICK TIPS

TIP #1: Determine the individuals/teams that are key to your organization hitting their goals. They will have a matrixed relationship and are vital to your success. Understand the relationship and set up a rhythm of connection with them for ongoing engagement. Sometimes, that engagement is just awareness.

6

BRINGING IT
ALL TOGETHER

e e ꝑ

How do you kickstart your chief of staff journey, or if you are on one, how do you audit and evaluate where things are to increase skilling? Here is a list of things you can do now. We will return to the executive pillars to help guide you. You need to focus on things you want to accomplish within three months with a target of creating a one-year+ strategy as an output from what you learn in those initial ninety days. This plan will work whether you are a new or an existing CoS.

"

This CoS team is the foundation
the organization is built upon.

As you facilitate the executives' needs and how you will drive the organization, you should also focus on the CoS strategy. What are you looking to accomplish? What are the guiding principles for your team? You can also determine your current and future state by examining an organizational structure. As you evaluate the organization's needs, one focus is always around scaling for growth. Your team should match pace and optimize growth.

Included in the appendix is a QR code for reference material. In it, you will find an exanple of an operational priorities framework. This should be based on what you have learned, feedback received from the team and stakeholders, and an understanding of gaps.

Most chief-of-staff organizations have many functions, including but not limited to:

- Budget and Finance Director
- Communications Director
- Operations Director
- Community and Culture Director
- Technical Writer
- Business Manager
- Administrative Leader
- Strategy Manager

Once you have determined whether you will be a team of one or many, you will want to review the job description and what is needed for alignment. This CoS team is the foundation of the organization; therefore, it is imperative that you understand the organization's strategy and vision. Sometimes, these are documented in a shared location, uploaded to the internal website, or communicated to your team somehow. You will determine your priorities, and these priorities should line up with the vision, creating a line of sight to what you execute against and the strategy. Your work should enable the organization to keep the lights on and move the business forward.

So now that you understand the possibilities of where this role can take you and the skills needed to be successful, what does it look like? If you were to shadow a chief, what would you see? Your entry in this journey will look much different than that of a seasoned veteran. As with all jobs, there is a ramp-up window or learning curve. for those starting a new role. You must determine if your role was newly created or if you are replacing an existing chief. If an existing chief existed, you would want to leverage the work done and evaluate what is needed and not already in place.

The first order of business is to understand the problems or issues that must be tackled immediately. As you start your week, you will need to evaluate the hot items that must be tackled. Were there problems or issues that required triaging? Are there key people you need to engage with, and does your executive or anyone from the leadership team need to be informed? The issues or problems you are addressing will be items that will require some digging on your part. You will be collaborating with a variety of different individuals. You will be tackling this problem and trying to understand if there's a deadline and when you need to have it resolved. That will take up some part of your day. Another thing that will happen is that you will also ensure that you have essential one-on-ones (1:1) scheduled and a review of your week with what is going on.

I typically start my day with a connection to my executive and a review of the tasks that are in progress, the hot things that I need to work on, and ask for anything that I need to have on my radar for the week to support or help that executive or our team. As you review your week, you must allocate time to focus, setting aside uninterrupted time to work on projects, answer questions, or review emails. You'll also want to include time for a break if you have back-to-back meetings, allowing yourself to decompress or walk around the office, home, or outside. You will also want to add time for lunch. It sounds silly, but sometimes, if our days are full, we might forget to eat and refuel.

If you have staff, this may also be an excellent opportunity to connect with them to align with what is expected for the week and if you have things that will spread into the following week or weeks. Many organizations will schedule their team or staff meetings early in the week again as a look forward and address things that require attention sooner rather than later. Learning and development are fundamental, so you'll want to add time.

Whether understanding the latest technology your team will be working on you can provide books to read. You may choose some time to invest in sitting down and just reading. This is a good time for you to do some data analysis or tactical work for any upcoming executive reviews on the calendar. You may require information from different members of the team that you can start to solicit and pull together. After a while, you will see that with a handful of meetings and a few blocks of focus time, your day will be well spent with self-driven activities supporting your organization's needs. Depending on your scope, you can drive your workday.

Some crucial conversations you should have should include mentors and sponsors who can provide you with perspective and an opportunity to see things from a different vantage point as you work.

In a time when there's so much hybrid work, most roles in technology do not happen in a traditional 8 to 5 time zone. You may have team members or leaders who live in different parts of the world and different time zones. You also find that you have times that are more intense and might require a twelve-hour day and times where less is required. Therefore, regular scrutiny of where your time is spent and how it's spent will help you achieve a level of balance as you tackle the many facets of a chief of staff role—allowing you to balance work, home life, and your development.

The role of a chief of staff has emerged as a catalyst for organizational success. As the link between the technical and executive teams, this position requires a unique set of skills and a strong understanding of both the workings of technology and the broader strategic goals of the company or organization. We reviewed a few of the responsibilities, challenges, and benefits of being a chief of staff. We explored how this role contributes to organizational success and is a valued executive team member.

The chief is the principal person to the executive they partner with. They are strategic advisors. This role will need to balance operational and leadership expertise. We also reviewed how some responsibilities may include the following:

- **Strategic Alignment:** The chief of staff's key responsibility is aligning technical initiatives with the overall business strategy. This involves translating business goals into actionable plans and ensuring the roadmap aligns with the organization's vision. Strategic alignment is crucial for fostering innovation and driving the organization forward.

- **Project Management:** The chief of staff plays a critical role in overseeing projects and initiatives. This includes coordinating multiple teams, managing timelines, and ensuring that initiatives are completed and delivered in line with organizational objectives. Effective project management is essential for providing quality services on time and within budget.

- **Communication Liaison:** Bridging the gap between the various teams within and external to their respective team and executive leadership, the chief of staff must master their communication ability. Effective communication promotes collaboration and transparency. It also ensures that everyone is working towards a common goal.

- **Operational Efficiency:** Beyond strategy and projects, the chief of staff is tasked with optimizing operational processes to make sure the team is running effectively. This could involve streamlining workflows and keeping and executing against a communication plan. The chief is also proactive in identifying and mitigating issues, which includes establishing processes and procedures to create clarity and minimize confusion. This is critical for maintaining an edge in the fast-paced tech landscape.

The benefits and opportunities for development will seem endless. One of the most rewarding aspects of being a CoS is the opportunity to make a meaningful impact on the organization—a pivotal role in shaping the company's future by aligning the various initiatives with business goals and driving key projects to success. Being a chief can be a vehicle for executive growth, gaining exposure to both technical and executive leaders, and acquiring a diverse skill set that is highly valued in executive roles. This experience serves as a stepping stone for career advancement within the organization.

Successfully navigating the role's challenges often results in empowered and high-performing teams. By providing clear and understandable communication, strategic guidance, and efficient processes, the chief of staff contributes to a positive and productive work environment, fostering innovation and collaboration.

QUICK TIPS

TIP #1: Here is a checklist of ten immediate items you can start with as you build your onboarding plan. As you drive your team, you will want to adapt to your specific business needs.

- Create a Kickoff Plan and milestones.
- Review strategic documents and establish priorities.
- There may be a period of education about the chief of staff, evangelizing, and getting support.
- If you have a team, meet early and often.
- Understand the priorities of your leadership team.
- Draft the framework and operational principles.
- Identify gaps and challenges. What is keeping your executive up at night? Understand why and take it off their plate.
- Where is your potential gap, and how can you minimize that impact? Is there someone you can partner with to fulfill that need?
- What are the upcoming targets and reviews that your executive needs to prepare for?
- Which HOT KPIs are not tracking to success? Work with that leader or team member to help understand what steps need to be taken to get back on track?

APPENDIX

To download complimentary reference
material, scan the QR code below:

KEY DEFINITIONS

FY: Fiscal Year - A fiscal year is an annual accounting period of 12 months for which a government, business, or other organization reports its financial information. A fiscal year does not necessarily coincide with the calendar year and may vary depending on the revenue cycle of the entity.

CY: Calendar year - A calendar year is a one-year period that begins on January 1 and ends on December 31, based on the Gregorian calendar.

OKR: Objective and Key Results - Alternatively OKRs, Objective and Key Results is a goal-setting framework used by individuals, teams, and organizations to define measurable goals and track their outcomes. The development of OKR is generally attributed to Andrew Grove who introduced the approach to Intel in the 1970s.

KPI: Key Performance Indicators - A Key Performance Indicator (KPI) is a measurable target that indicates how individuals or businesses are performing in terms of meeting their goals and defined strategy.

MBO: Management by Objectives - Also known as management by planning (MBP), Management by Objectives was first popularized by Peter Drucker in his 1954 book *The Practice of Management.* Management by objectives is the process of defining specific objectives within an organization that management can convey to organization members, then deciding how to achieve each objective in sequence.

GDPR: General Data Protection Regulation compliance is the adherence to the data protection principles regarding personal data of EU citizens by any organization that acts as a data controller or handles such data. GDPR compliance is not optional and can result in severe fines for violations. To demonstrate GDPR compliance, organizations should conduct a data protection impact assessment and always take data protection into account.

Matrixed organizations: an employee who may not have a direct reporting relationship with a manager but has functional relationship.

RESOURCES

Six Sigma: The Council for Six Sigma Certification - Official Industry Standard (sixsigmacouncil.org)

Six Sigma is a set of techniques and tools for process improvement that was introduced by American engineer Bill Smith while working at Motorola in 1986[1]. The methodology seeks to improve manufacturing quality by identifying and removing the causes of defects and minimizing variability in manufacturing and business processes[1]. The name "Six Sigma" comes from statistical quality control, which evaluates process capability. It refers to the ability of manufacturing processes to produce a very high proportion of output within specification.

The Six Sigma methodology is based on a five-phase approach called DMAIC, which stands for Define, Measure, Analyze, Improve, and Control[2]. The goal of DMAIC is to identify and eliminate the root causes of defects and minimize variability in business and manufacturing processes. The Six Sigma methodology has been widely adopted by many organizations around the world, including Fortune 500 companies. It has been used to improve quality, reduce costs, increase customer satisfaction, and boost profits.

Project Management Professional : <u>Project Management Institute</u> <u>| PMI</u>

PMP stands for Project Management Professional. It is an internationally recognized professional designation offered by the Project Management Institute (PMI) . The PMP acknowledges candidates skilled at managing the people, processes, and business priorities of professional projects . PMI, the world's leading authority on project management, created the PMP to recognize project managers who have proven they have project leadership experience and expertise in any way of working . To obtain PMP certification, a project manager must meet certain requirements and then pass a 180-question exam . The PMP exam was created by project executives for project executives, so each test question can be related to real-life project management experiences.

Scrum : <u>Scrum Alliance Certification | Transform your workplace</u>

Scrum is the Framework in which a sprint takes place. A Sprint is a defined time period for developing features for a product. The maximum time for a sprint is 30 days (can be shorter but not longer). During a sprint the development team develops new features for the product. When the sprint is finished a new version of the product is available. This product could be shipped to the customer.

The Information Technology Infrastructure Library (ITIL) is a set of detailed practices for IT activities such as IT service management (ITSM) and IT asset management (ITAM) that focus on aligning IT services with the needs of the business[1]. ITIL describes processes, procedures, tasks, and checklists which are neither organization-specific nor technology-specific but can be applied by an organization toward strategy, delivering value, and maintaining a minimum level of competency[1]. It allows the organization to establish a baseline from which it can plan, implement, and measure. It is used to demonstrate compliance and to measure improvement[1]. There is no formal independent third-party compliance assessment available for ITIL compliance in an organization. Certification in ITIL is only available to individuals[1].

PRINCE2 is a process-based method for effective project management, and will give you the fundamental skills you need to become a successful project manager. It stands for **PR**ojects **IN** **C**ontrolled **E**nv ironments, and is used and recognized all over the world. PRINCE2 is completely scalable, and the most recent update to the framework means it can be more easily tailored to each project you undertake. PRINCE2® has recently been updated from the 6th Edition to the 7th Edition, and the PRINCE2 7th Edition Foundation and Practitioner exams will be available for individuals and organizations from September 2023.

Management by objectives (MBO) is a management framework whose goal is to push higher performance of an organization by clearly defining objectives that are agreed to by both management and employees. As a person on a team or a team, if your voice is head is creating and setting to objectives or goal will have a greater chance of success because it encourages participation and commitment with te team, as well as aligning objectives across the organization.

Key Takeaways

- Management by objectives (MBO) is a process in which a manager and an employee agree on specific performance goals and then develop a plan to reach them.
- It is designed to align objectives throughout an organization and boost employee participation and commitment.

- There are five steps: Define objectives, share them with employees, encourage employees to participate, monitor progress, and finally, evaluate performance and reward achievements.

The OKR methodology was created by Andy Grove at Intel and taught to John Doerr by him. Explore the <u>complete OKR origin story</u> and learn more about John Doerr's *<u>Measure What Matters</u>. *

- In the book Measure What Matters, John Doerr writes about "MBOs," or Management by Objectives. MBOs were the brainchild of Peter Drucker and provided Andy Grove a basis for his eventual theory of OKRs. In fact, Grove's name for them originally was "iMBOs," for Intel Management by Objectives. Despite the original name, Grove created some key differences between the two which he passed along to Doerr.
- Grove rarely mentioned Objectives without tying them to "Key Results," a term he seems to have coined himself. Other key differences between MBOs and OKRs are that the latter are quarterly, not annual, and they are divorced from compensation.
- Doerr was the one who crafted the name "OKRs." He introduced the philosophy to Google's founders in 1999. Gathered around a ping-pong table which doubled as a boardroom table, Doerr presented a PowerPoint to the young founding team, which included Larry Page, Sergey Brin, Marissa Mayer, Susan Wojcicki, and Salar Kamangar

Contact Me!

ownyourpowerwithannemarie

annemarieotanez

ownyourpowerwannemarie

OwnYourPowerwithAnneMarie

www.annemarieotanez.com

Made in United States
Orlando, FL
15 September 2024